高等学校交通运输与工程类专业教材建设委员会规划教材

CONCISE SOIL MECHANICS

简明土力学

(English - Chinese)

(英汉对照)

Lead authors Yue Zurun Hu Tianfei
Lead reviewer Sun Tiecheng

主 编 岳祖润 胡田飞
主 审 孙铁成

人民交通出版社
北 京

Abstract

内 容 提 要

This book is compiled according to the domestic soil mechanics syllabus and the British undergraduate teaching materials, and has been used and revised by several undergraduate courses. Main features: First, English-Chinese comparison is convenient for professional vocabulary learning and teaching of foreign students; Second, it is concise, highlighting the basic concepts and principles; The third is to straighten out the logical relationship of knowledge points according to cognitive rules, so that students can grasp the course as a whole.

本书根据国内土力学教学大纲和英国本科教材进行编译,经过多届本科生教学使用并修订而成。本书主要特点:一是英汉对照,便于专业词汇学习和留学生教学;二是内容层次简明,突出基本概念和基本原理;三是按认知规律理顺知识点间逻辑关系,便于学生从整体上把握土力学课程的知识体系。

This book is divided into 10 chapters, mainly including the physical properties of soils, stress distribution in soils, flow of water through soils, consolidation and settlement of soils, shear strength of soils, lateral earth pressure, slope stability, foundation bearing capacity, engineering geological investigation, and intelligent compaction. Each chapter is accompanied by exercises at the end of the book for students' practice.

本书共分 10 章,主要包括:土的物理性质、土中应力分布、土的渗流、土的固结与沉降、土的抗剪强度、土压力、边坡稳定性、地基承载力、工程地质勘察和智能压实。每章附有习题,供学生练习。

This book can be used as a bilingual course textbook of soil mechanics for undergraduate students of civil engineering and related majors in colleges and universities and a textbook for foreign students. It can also be used as a reference book for relevant scientific and technological personnel.

本书可作为高等院校土木工程及相关专业本科生土力学双语课程教材和留学生教材,也可作为相关科技人员的学习参考资料。

Preface
前　　言

The fragmentation, heterogeneity, and variability of soil make the study of soil mechanics different from other mechanical courses. The theory of soil mechanics needs to be based on experimental research. Currently, the deformation and strength characteristics of soil are studied separately, making it difficult for students to grasp the relationship between soil deformation and strength characteristics, leading to confusion when learning soil mechanics.

土体的碎散性、多相性和变异性使土力学与其他力学课程有一定区别。土力学的理论需要以试验研究为基础,目前土体的变形特性和强度特性是分开研究的,使学生很难掌握土体变形特性和强度特性之间的联系,因而学生在学习土力学时会产生诸多困惑。

Based on years of teaching practice, we try to organize the logical relationships between knowledge points, which plays a crucial role in the learning process. This textbook introduces the physical properties of soil (material composition, structure, three-phase indicators, state indicators, and engineering classification) and stress calculation (self weight stress and additional stress) firstly; Then, we analyse the permeability and deformation characteristics of the soil, and based on Darcy's law and Terzaghi's consolidation theory, the permeability, permeability stability, consolidation and settlement characteristics of the soil were studied; Afterwards, we study the strength and stability of soil, and provide the calculation methods for soil pressure, slope stability, and foundation bearing capacity based on the Mohr-Coulomb strength criterion; Finally, a brief introduction was given to engineering geological survey and compaction methods of soil, enabling students to understand the application and development of soil mechanics. These two parts can be simplified or detailed according to the class hours. The first feature of this textbook is to sort out the logical relationship between knowledge points.

结合多年的教学实践,我们试着整理出土力学知识点间的逻辑关系,这在指导学生学习过程中能起到纲举目张的作用。本教材首先介绍了土的物理性质(物质组成、结构、三相指标、状态指标和工程分类)和应力计算(自重应力和附加应力);然后分析了土的渗透和变形特性,介绍了达西定律和太沙基固结理论,研究了土的渗透性、渗透稳定性、固结和沉降特性;之后研究了土的强度和稳定问题,介绍了莫尔-库仑强度准则,土压力、边坡稳定和地基承载力计算方法;最后简略介绍了工程地质勘察和土的压实方法,使学生了解土力学的应用和发展,这两部分内容根据课时可简可繁。理出知识点间的逻辑关系是本教材的第一个特色。

The importance of soil mechanics in the study of civil engineering is unquestionable. Due to limited class hours, mastering the essence of courses such as basic concepts and principles has become very important. This textbook not only focuses on extracting key knowledge points, but also ensures the systematicity and completeness of knowledge. For example, in the introduction of slope stability and foundation bearing capacity calculation methods, not all the theories proposed by scientists were listed, only representative theories were selected, and other related topics were further studied according to individual needs. The second characteristic of this textbook is its concise content and prominent focus.

土力学对土木工程专业后续专业课学习的重要性毋庸置疑。由于学时有限,对于基本概念和基本原理等课程精髓的掌握变得尤为重要。本教材在注重提炼关键知识点的同时,保证了知识的系统性和完整性。例如,在边坡稳定和地基承载力计算方法介绍中,没有罗列所有科学家提出的理论,只挑选具有代表性的理论,其他相关专题视个人需要再深入研究。内容简明、重点突出是本教材的第二个特色。

Authentic professional English should be learned in conjunction with professional courses, and bilingual English-Chinese textbooks are the preferred choice for professional English learning. In the process of learning professional knowledge, it can not only improve English reading and writing skills, but also deepen understanding of problems, getting a double advantage, which is the third feature of this textbook.

地道的专业英语应当结合专业课进行学习,英汉对照双语教材是专业英语学习的首选。在学习专业知识的过程中,既可提高英文阅读和写作水平,又可加深对问题的理解,一举两得,这是本教材的第三个特色。

This textbook was written by Yue Zurun and Hu Tianfei from ShijiazhuangTiedao University, and reviewed by Sun Tiecheng. During the writing process, a large number of excellent domestic and foreign textbooks were referenced. We would like to pay tribute to Chen Zhongyi, Deng Longji, Liang Zhongqi, Liu Chengyu, and other leaders in the field. During the writing process, doctoral students Li Congran and Mao Shuai, and master's students Gu Xiangtao, Zhang Yimin and Li Xiaoxing have enthusiastically done a lot of work, and I would like to express my gratitude to them.

本教材由石家庄铁道大学岳祖润、胡田飞编写，孙铁成主审。编写过程中参考了大量的国内外优秀教材。在此向陈仲颐、邓隆基、梁仲琪、刘成宇等老前辈致敬！在编写过程中，博士研究生李聪然、毛帅，硕士研究生顾相涛、张翊敏、李晓星热心地做了大量工作，在此谨向他们表示感谢。

Editor

编　者

January 2025

2025 年 1 月

Logical Relationship of Knowledge Points

土力学知识点逻辑关系

- Soil Mechanics 土力学
  - Physical Properties 土的物理性质
    - Index of Phases 三相的数学指标
    - Index of Physical State 物理状态指标
    - Classification of Soil 土的分类
  - Loading Properties 土的荷载性质
    - Self Weight Stress 自重应力
    - Additional Stress 附加应力
  - Seepage and Deformation 土的渗透与变形 ⇒ Darcy Theory, Terzaghi Theory 达西定律、太沙基理论 ⇒
    - Permeability 渗透性
    - Consolidation 固结
    - Final Settlements 最终沉降量
  - Strength and Stability 土的强度与稳定 ⇒ Mohr-Coulomb Theory 莫尔-库仑理论 ⇒
    - Soil Pressure 土压力
    - Slope Stability 边坡稳定
    - Bearing Capacity 地基承载力
  - Engineering Investigation and Compaction of Soil 工程地质勘察与土的压实

Logical Relationship of Knowledge Points

土力学知识点逻辑关系

- Soil Mechanics 土力学
  - Physical Properties 土的物理性质
    - Index of Phases 三相的数量指标
    - Index of Physical State 物理状态指标
    - Classification of Soil 土的分类
  - Loading Properties 土的应力性质
    - Self-Weight Stress 自重应力
    - Additional Stress 附加应力
  - Seepage and Deformation 土的渗透与变形
    - Darcy Theory, Terzaghi Theory 达西定律、太沙基理论
      - Permeability 渗透性
      - Consolidation 固结
      - Final Settlement 最终沉降
  - Strength and Stability 土的强度与稳定
    - Mohr-Coulomb Theory 莫尔-库仑理论
      - Soil Pressure 土压力
      - Slope Stability 边坡稳定
      - Bearing Capacity 地基承载力
  - Engineering Investigation and Compaction of Soil 工程地质勘察与土的压实

Contents
目　　录

Introduction of Soil Mechanics/土力学概论

Chapter 1　Physical Properties of Soils/土的物理性质

Chapter 2　Stress Distribution in Soils/土中应力分布

Chapter 3 Flow of Water through Soils/土的渗流

Chapter 4 Consolidation and Settlement of Soils/土的固结与沉降

Chapter 5 Shear Strength of Soils/土的抗剪强度

Chapter 6 Lateral Earth Pressure/土压力

Chapter 7 Slope Stability/边坡稳定性

Chapter 8 Foundation Bearing Capacity/地基承载力

Chapter 9 Engineering Geological Investigation/工程地质勘察

Chapter 10 Intelligent Compaction/智能压实

References/参考文献

Introduction of Soil Mechanics
土力学概论

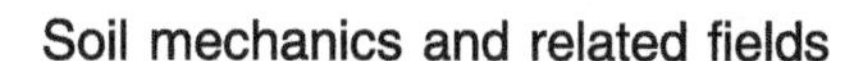

Soil mechanics and related fields

Soil mechanics is one of the engineering disciplines that deal with soils as an engineering material. Since ancient ages, engineers have been handling soils as an engineering material for various construction projects and constructed a large amount of such magnificent achievements, such as the Egyptian pyramids, Mesopotamian ziggurats, Roman aqueducts, and China's Great Wall. However, those ancient projects were mostly accomplished by accumulated experiences of engineers. During the eighteenth and nineteenth centuries, some modern engineering theories were employed in this field, following the development of Newtonian mechanics. Coulomb's and Rankine's lateral earth pressure theories (Chapter 6) are some examples of such theories.

The modern era of soil mechanics had to wait until 1925, when Dr. Karl Terzaghi published a book called *Erdbaumechanik*. His then-new concept of "effective stress", which deals with interaction with pore water, has revolutionized the mechanics of soils. The development of modern soil mechanics is due to his great contribution. Karl Terzaghi is now regarded as the father of modern soil mechanics.

Soil mechanics contains foundation engineering, geotechnical engineering, and geoenvironmental engineering. Foundation

土力学及相关领域

土力学是将土作为工程材料的学科。自古以来,工程师们将土作为主要建筑材料,建成了大量宏伟的工程,如埃及金字塔、美索不达米亚塔庙、罗马渡槽和中国长城。这些工程大多是根据工匠们积累的经验完成的。在18~19世纪,随着牛顿力学的发展,一些现代工程理论被应用于土力学领域,如库仑和朗肯的侧向土压力理论(第6章)。

现代土力学起始于1925年,卡尔·太沙基(Karl Terzaghi)博士出版了一本名为《土力学》(*Erdbaumechanik*)的著作,提出了"有效应力"的概念,反映土与孔隙水的相互作用,这彻底改变了土力学。现代土力学的发展得益于他的巨大贡献,因此太沙基被视为"现代土力学之父"。

土力学包括基础工程、岩土工程和环境岩土工程。基础工程是根据土力学知识设计建筑基础、挡土结构、路堤、土石坝和边坡等结构的学科。多年来该学科一直被称为土力学与基础工程。岩土工程这一新术语是在1970年左右提出的，目的是将岩石力学、土力学及基础工程相结合。在20世纪80年代，与环境相关的岩土工程备受关注，此时提出了“环境岩土工程”一词，该工程包括固体及液体废弃物处置设施的设计和施工，以及其他与环境相关的岩土工程问题。

engineering is the field of designing foundations of buildings, retaining structures, embankments, earth and rockfill dams, and safe earth slopes, etc, based on the knowledge of soil mechanics. Thus, the discipline has been called soil mechanics and foundation engineering for many years. The new term, geotechnical engineering, was coined around 1970 to merge rock mechanics into soil mechanics and foundation engineering. In the 1980s. environmentally related geotechnical engineering became a great engineering concern, and the term geoenvironmental engineering was created. This includes the design and construction of solid- and liquid-waste containment facilities and any other environmentally related geotechnical engineering problems.

卡尔·太沙基(Karl Terzaghi)的博士传记

1883年，卡尔·太沙基博士出生于奥地利布拉格。1900年，他进入格拉茨工业大学(Graz University of Technology)学习机械工程。1904年，他以优异成绩毕业。1925年，他出版了《理论土力学》和《实用土力学》，这两本书彻底改变了土力学的发展，同年太沙基获得了麻省理工学院(Massachusetts Institute of Technology, MIT)的职位。1936年，太沙基在哈佛大学(Harvard University)举办的第一届土力学与基础工程国际会议上做了演讲。1936—1957年，他担任国际土力学与基础工程学会首任主席。为了表彰太沙基在土力学领域的巨大贡献，美国土木工程师协会(American Society of Civil Engineers, ASCE)于1960年设立了卡尔·太沙基奖。太沙基的贡献在于提出了有效应力、固结等理论以及抗剪强度和承载力等计算方法。太沙基于1963年去世。

Biography of Dr. Karl Terzaghi

Dr. Karl Terzaghi was born in Prague, Austria, in 1883. He entered the Graz University of Technology to study mechanical engineering in 1900. He graduated with honors in 1904. In 1925, he published *Theoretical Soil Mechanics* and *Practical Soil Mechanics*, which revolutionized the field to great acclaim and resulted in the offer of a position from the Massachusetts Institute of Technology (MIT). Terzaghi gave a plenary lecture at the First International Conference on Soil Mechanics and Foundation Engineering at Harvard University in 1936. He served as the first president of the International Society of Soil Mechanics and Foundation Engineering from 1936 to 1957. In honor of his great contribution in soil mechanics field, the American Society of Civil Engineers (ASCE) established the Karl Terzaghi Award in 1960. Terzaghi's contribution include effective stress, consolidation, shear strength and bearing capacity theory, etc. Terzaghi died in 1963.

Uniqueness of soils

Soil is a very unique material and complex in nature. The unique characteristics of soils are as follows:

(1) Soil is not a continuous solid material, but rather is composed of three different constituents: solid particles, water and air.

(2) Particle sizes have significant influence on soil behavior.

(3) The amount of water also plays a very important role in soil behavior.

(4) The stress strain relation of soil is nonlinear from the small strain levels.

(5) The pore spaces of soil possess the capability of water flow.

(6) Soil is susceptible to creep.

(7) Soil swells when wetted or shrinks when dried.

(8) Soil is an anisotropic material due to the particle shapes and the depositional direction under gravity.

(9) Soil is also spatially nonhomogeneous.

To describe these unique nature, the soil mechanics utilizes many different areas of mechanics. For the various phases, researches are conducted by solid mechanics as well as discrete mechanics. The water flow characteristics are explained by knowledge of fluid mechanics such as Darcy's law and Bernoulli's law. Physicochemical knowledge is required to understand swell and shrinkage characteristics. Understan-ding its anisotropic characteristics requires a high level of knowledge in mechanics and material science. Some statistical approaches are also needed to treat the non-homogeneity of soils.

As briefly seen earlier, soil is a unique material, and its engineering properties vary a lot depending on the particle sizes, origins and many other factors. Their constitutive models are not as simple as Hooke's law which is used in some other materials.

土的独特性

土是一种特殊材料，其特性在自然界中非常复杂。土的特性如下：

(1)土体不是连续的固体材料，是由三种成分组成(固体颗粒、水和空气)的材料。

(2)颗粒大小对土体特性有显著影响。

(3)含水率对土体特性有显著影响。

(4)在小应变水平上，土体应力-应变关系是非线性的。

(5)水可以在土体孔隙间流动。

(6)土体具有蠕变特性。

(7)土体具有湿润时膨胀、干燥时收缩的特性。

(8)土体的颗粒形状和重力作用下的沉积方向不同，使土体成为一种各向异性材料。

(9)土体在空间上不均匀。

为了描述这些特性，土力学运用了不同力学理论。对于不同的相态，运用固体力学和散体力学。进行研究对于土的渗流特性，运用流体力学中的达西定律和伯努利定律进行解释；对于膨胀和收缩特性，运用物理和化学理论进行解释；对于各向异性，运用高等力学理论和材料学理论进行解释；对于非均匀性，运用统计方法进行分析。

如前所述，土是一种特殊材料，其工程性质因颗粒大小、形成原因及其他因素而异。土体的本构模型也不像一些其他材料使用的胡克定律那样简单。

土力学问题的探讨

由于土的复杂性和空间变异性,对其开展原位观测和室内试验非常重要。原位观测主要包括现场地质调查和土样采集,有时还包括土体性质的现场测试,例如通过井孔抽水试验测定渗透性、通过十字板剪切试验测定强度等。采集的土样可在实验室用来进行各种物理和力学试验,物理性质试验包括颗粒级配试验、界限含水率试验、相对密度试验和渗透试验等,力学试验包括击实试验、固结试验和各种强度试验等。

根据原位观测和室内试验数据,结合现代土力学知识和基础工程概念,对土进行分类,确定设计指标及安全合理的基础和土工结构设计方案,施工单位根据设计方案进行施工。通常,设计工程师会仔细监控施工过程,确保工程项目按照设计方案正确执行。

Approaches to soil mechanics problems

Complexity and spatial variation of soil make the field observation and laboratory testing necessary. Field observation mainly ranges from geological study of the site to soil sampling and sometimes in-sit testing of properties, such as well tests for permeability, vane shear tests for strength determination, etc. Sampled specimens are brought back to laboratories for various physical and mechanical tests. The former includes the grain size test, atterberg limits tests, specific gravity test, and permeability test etc, and the latter includes a compaction test, consolidation test, and various shear strength tests.

Based on field observations and laboratory test data, geotechnical engineers classify soils, determine design properties, and design safe foundations and earth structures, by fully utilizing modern soil mechanics knowledge and foundation engineering concepts. Construction companies carry out construction of the project according to design. Usually, design engineers monitor construction practices carefully for proper execution.

Physical Properties of Soils 土的物理性质 Chapter 1

1.1 Composition of soils

Soil can be defined as an assemblage of solid particles, and it consists of three phases: solid, liquid (water), and gas (air). Gravel, sand, silt, and clay are commonly used names of soils of different particle grain sizes.

The rock cycle shown in Figure 1-1 illustrates the origins of a variety of soils on the Earth. Most original rock is formed from molten magma deep in the Earth. The next process is weathering. Soils are formed from the physical and chemical weathering of rocks. Deposited soils are often subjected to many geological years of cementation and compression that transform them into sedimentary rocks such as sandstone, shale, limestone, dolomite, and many others. Rocks can undergo further transformation in response to exposure to high heat and pressure in deeper Earth but without melting. This process is called metamorphism. Metamorphic rocks can melt into magma in the deeper layers of Earth to complete the rock cycle (Figure 1-1). Sedimentary rocks and metamorphic rocks are also subjected to weathering, transportation, and deposition processes that result in the formation of soils.

(1) Physical weathering involves a reduction in size without any change in the original composition of the parent rock. This process mainly occurs due to exfoliation, unloading, erosion, freezing, and thawing.

1.1 土的构成

土是各类固体颗粒的集合体,是由固体、液体(水)和气体(空气)组成的三相体。根据土颗粒的粒径分类,常见的土类名称有砾石土、砂土、粉土和黏土。

图 1-1 中的岩石循环圈展示了地球上各种土的起源。大多数岩石由地球深处的熔融岩浆冷却形成,土是岩石物理风化和化学风化的产物。土颗粒经过风化、搬运、沉积等成岩作用,转化为沉积岩,如砂岩、页岩、白云岩和石灰岩等。沉积岩受到地球深处高温和高压引起的变质作用,形成变质岩。变质岩受到岩浆作用,熔化成地球深部的岩浆(图 1-1)。沉积岩和变质岩也会受到风化、搬运和沉积作用,从而再次形成土。

(1)物理风化指由于剥落、卸载、侵蚀、冻结和解冻等因素的作用,母岩逐渐变为小尺寸的碎块和细小颗粒,但仍保持母岩的矿物成分。

(2)化学风化会改变母岩尺寸和矿物化学成分,其主要通过水解、碳化和氧化等方式进行。

(2) Chemical weathering causes both reductions in size and chemical alteration of the original parent rock. It mainly occurs due to hydration, carbonation, and oxidation.

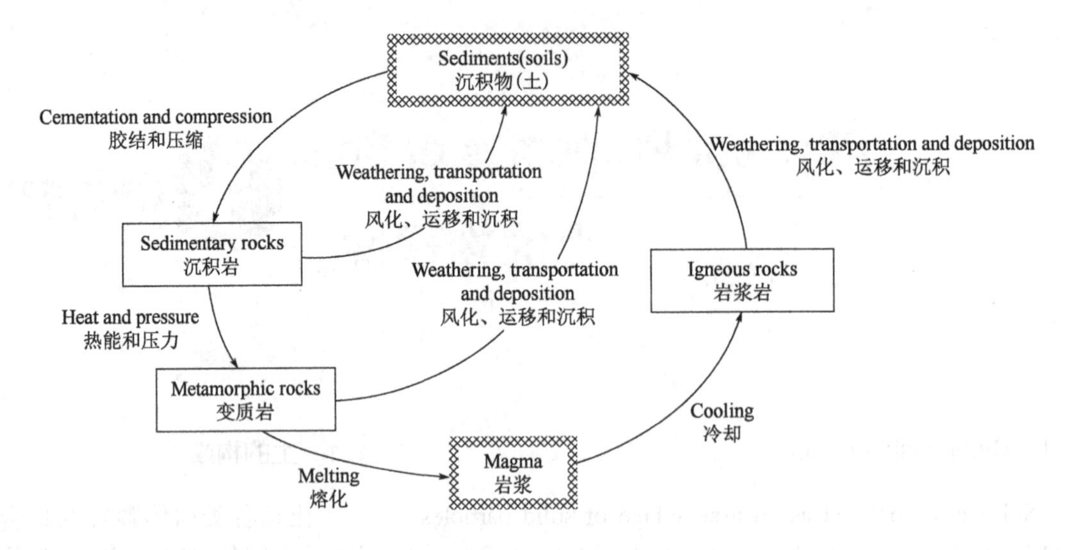

Figure 1-1 Rock cycle

图 1-1 岩石的循环

在自然界,化学风化和物理风化往往同时进行。岩石风化后,部分碎屑经水流、冰川和风等的搬运,会逐渐沉积在离原始位置不同距离的地方。但仍有部分碎屑会留在原地,形成残积土,其保留了母岩的许多元素。岩石碎屑在搬运过程中进一步受到各类风化作用,粒径会变得更小,形状更为圆滑,棱角更少。

Chemical and physical weathering often occur simultaneously. The next process in the cycle is transportation. Broken fragments of rock are displaced by running water, movement of glaciers, and sometimes by wind, and they eventually come to rest at a certain distance from the original location (deposition). Soils that remain at the site of weathering are called residual soils. These soils retain many of thc clements of the parent rock. During the transportation process, particles undergo further physical and chemical changes and thus become smaller and rounder.

岩石经过风化过程会形成不同类型的土,如不同的颗粒大小、形状和矿物成分等。目前,有许多关于土的描述和分类方法,列举部分如下。

The processes mentioned result in types of soil that are different in terms of their particle sizes, shapes, mineral compositions, and other properties. Some of the commonly used descriptions of soil types are listed below.

(1)冲积土是岩石风化后被江河的水流搬运并沉积在河谷上的细小沉积物,具有明显的层理构造,大部分人类工程建设都在冲积土层上进行。

(1) Alluvial soils are fine sediments that have been eroded from rock and transported by water and have settled on river and stream beds. The profiles of alluvial soils usually consist of layers of different soils. A considerable amount of construction activity occurs in and on alluvial soils.

(2) Colluvial soils (colluvium) are soils found at the bases of mountains that have been eroded by the combination of water and gravity.

(2)坡积土是由于重力、流水的冲刷剥蚀作用沉积在山麓上的沉积物。

(3) Eolian soils are sand-sized particles deposited by wind.

(3)风积土是由风力搬运形成的沙粒大小的颗粒沉积物。

(4) Expansive soils are clays that undergo large volume changes in response to cycles of wetting and drying.

(4)膨胀土是在干湿循环中体积发生显著变化的一类黏性土。

(5) Glacial soils are mixed soils consisting of rock debris, sand, silt, clay, and boulders.

(5)冰碛土是由岩屑、砂土、粉土、黏土和巨砾组成的混合土。

(6) Lacustrine soils are mostly silts and clays deposited in lake waters.

(6)湖泊沉积土是沉积在湖水中的粉土和黏土。

(7) Lateritic soils are residual soils that are cemented with iron oxides.

(7)红土是由富铁氧化物胶结的残积土。

(8) Loesses are uniform, wind-blown, fine-grained soils.

(8)黄土是由风力作用搬运形成的细粒土。

(9) Marine soils are sand, silt, and clay deposited in salt or brackish water.

(9)海相沉积土指沉积在盐水或微咸水中的砂土、粉土及黏土。

(10) Mud is clay and silt mixed with water into a viscous soil.

(10)淤泥是由黏土、粉土与水混合而成的黏性软土。

1.2 Structure of soils

1.2 土的结构

Soil particles are assumed to be rigid. During deposition, mineral particles are arranged into structural frameworks called the soil fabric.

Two common types of soil fabric, flocculated and dispersed, are formed during the deposition of fine-grained soils, as shown schematically in Figure 1-2. In a saltwater environment, a flocculated structure is formed when many particles tend to orient themselves parallel to one another [Figure 1-2a)], whereas in a freshwater environment, such a structure is formed when many particles tend to orient themselves perpendicular to one another [Figure 1-2b)]. A dispersed structure is formed when most of the particles orient themselves parallel to one another [Figure 1-2c)].

假设土颗粒是刚性体,土颗粒在沉积过程中排列成的框架称为土的结构。

细粒土在沉积过程中,会形成两种常见的结构,如图 1-2 所示。在咸水环境中,絮凝结构的颗粒倾向于彼此平行,如图 1-2a)所示;在淡水环境中,絮凝结构的颗粒倾向于彼此垂直,如图 1-2b)所示;当大多数颗粒彼此平行时,就会出现分散结构,如图 1-2c)所示。

The spaces between the mineral particles, called voids, may be filled with liquids (essentially water), gases (essentially air), and cementitious materials (e.g., calcium carbonate). Voids typically occupy a large proportion of the soil volume. Interconnected voids form passageways through

矿物颗粒之间的空隙称为孔隙,可以被液体(主要是水)、气体(主要是空气)和胶结材料(例如碳酸钙)填充。孔隙通常占到土体积的很大一部分。相互连接的孔隙形成通道,水会通过这些通道进出土体。孔隙体积的变化导致土的压缩或膨胀。

which water flows in and out of soils. Changes in the volume of voids result in the compression or expansion of soil.

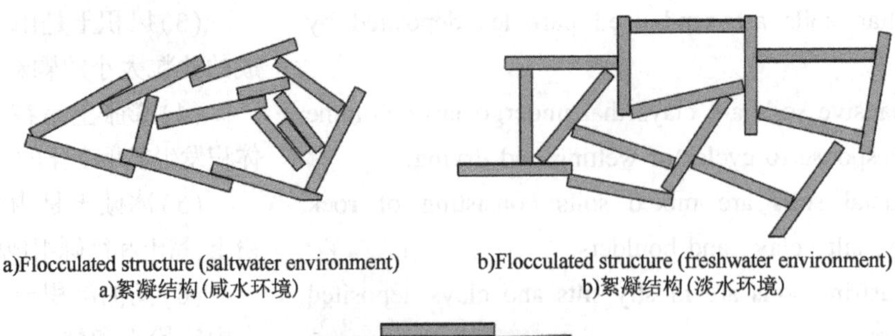
a)Flocculated structure (saltwater environment)
a)絮凝结构(咸水环境)
b)Flocculated structure (freshwater environment)
b)絮凝结构(淡水环境)

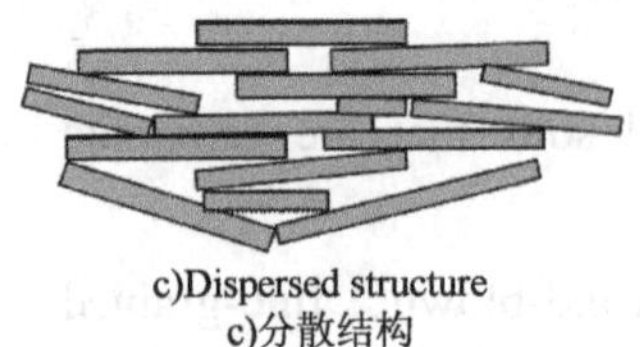
c)Dispersed structure
c)分散结构

Figure 1-2 Soil fabric
图 1-2 土的结构

如果用球体表示粗粒土中的刚性土颗粒(图 1-3),当球体逐一堆积在另一个球体上时,土体结构最为松散(孔隙空间最大),如图 1-3a)所示;当球体以交错方式排列时,土体结构最为密实,如图 1-3b)所示。实际上,粗粒土由各种粒径和形状的土颗粒组成,土颗粒的堆积形式是随机的。

If the rigid particles in coarse-grained soils can be approximated by spheres (Figure 1-3), then the loosest packing (corresponding to the maximum void space) occurs when the spheres are stacked on top of one another [Figure 1-3a)]. The densest packing occurs when the spheres are packed in a staggered pattern, as shown in Figure 1-3b). Real coarse-grained soils consist of an assortment of particle sizes and shapes, and consequently, the packing is random.

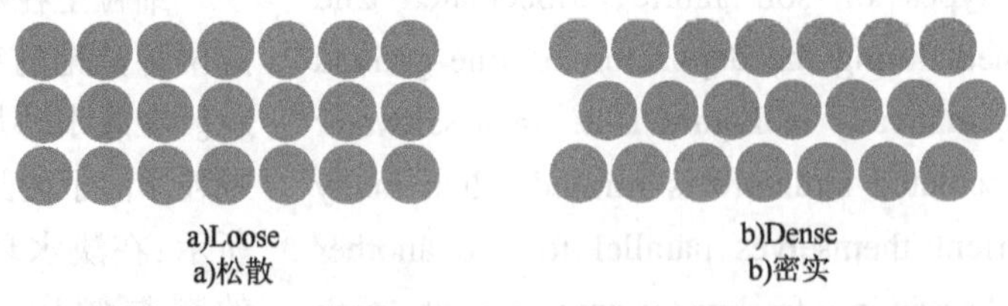
a)Loose
a)松散
b)Dense
b)密实

Figure 1-3 Loose and dense packing of spheres (coarse-grained soils)
图 1-3 松散和密实的球体排列方式(粗粒土)

1.3 土的粒径分析

土颗粒的大小是区分土体类型的重要依据。土体的常用名称如砾石土、砂土、粉土和黏土都是基于粒径划分的。粒组划分及各粒组对应的粒径范围如图 1-4 所示。砾石土和砂土视

1.3 Particle size analysis of soils

Particle size plays a dominant role in distinguishing soil types. Commonly used descriptors of soil, such as gravel, sand, silt, and clay, are based on their grain sizes. Figure 1-4 shows the ranges of grain sizes that correspond to those names. Gravel and sand are considered cohesionless (granular) soils, and clays are considered cohesive soils. Silt is a transitional material between granular soils and cohesive soils. These two soil groups can be distinguished by differences in their engineering properties and behavior. The resistance of granular soils to shearing derives mostly from surface friction and interlocking mechanisms. In contrast, the resistance of cohesive soils to shearing derives from particle-to-particle interaction. The former results in less compression than the latter and allows for much greater water flow through the particles.

为无黏性(粗粒)土,黏土视为黏性土,粉土是介于粗粒土和黏性土之间的过渡材料。这两类土可以根据其工程性质差异区分。粗粒土的抗剪强度主要由颗粒表面之间的摩擦作用和颗粒之间的咬合作用提供,而黏性土的抗剪强度由颗粒间的相互作用力提供。粗粒土的压缩性比黏性土小,但透水性更强。

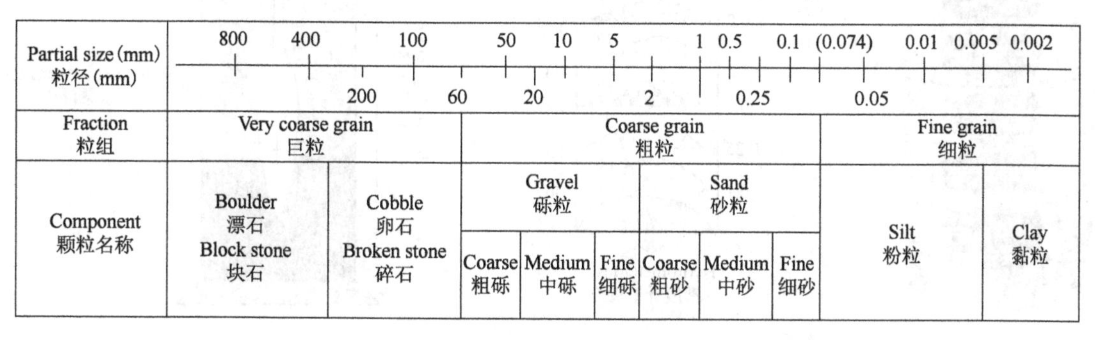

Figure 1-4 Soil types with respect to grain size ranges

图 1-4 土的粒组划分标准

The distributions of particle sizes and the average grain diameters of coarse-grained soil (gravels and sands) are obtained from sieve analysis. A typical stack of sieves, with various opening sizes (e. g., 0.075, 0.25, 0.5, 1, 2, 5, 10, 20, and 200 mm), is shown in Figure 1-5. The sieves are arranged based on the decreasing order of opening sizes from top to bottom. An oven-dried specimen of a known weight is placed on the top of the sieve stack, and a pan is placed on the bottom of the sieve stack. The stack of sieves is vibrated vertically and horizontally until no more weight change occurs in any of the sieves. A typical shaking period is approximately 8 to 10 min. After the shaking, the weights of the soils retained on each sieve are carefully measured on a balance to confirm that the initial total weight and the sum of the weights on each sieve after sieving are nearly equal.

粗粒土(砾石土和砂土)的颗粒级配或平均粒径通过筛分法确定。土的标准筛如图 1-5 所示,孔径分别为 0.075 mm、0.25 mm、0.5 mm、1 mm、2 mm、5 mm、10 mm、20 mm、200 mm,由上至下按照孔径递减依次排列,下附底盘。试验时将已知质量的风干土样放入最上层筛中,垂直和水平振动标准筛直到每个筛子内的土体重量不再变化,持续时间一般为 8 ~ 10 min。停机后,顺次将各筛上的土称重,并确认筛分前的质量和筛分后每个筛子上的质量总和相等。

The laboratory method commonly used to determine the grain size distribution of fine-grained soils is a hydrometer test, shown in Figure 1-6. Hydrometer analysis is performed for the material collected in the chassis in the sieve analysis. Distilled water is used to transfer the soil-water slurry to a

测量细粒土颗粒级配的常用方法是密度计法,如图 1-6 所示。将筛分试验中底盘收集的土料用蒸馏水散开制成悬液,然后放入密度计,测定不同时间密度计下沉的距离。根据土工试验手册中密度计法的试验步骤、理论和计算方法,计算土粒小于或等于某一粒径时的百分含量。

hydrometer cylinder to obtain a fully mixed uniform suspension. The exact times and corresponding hydrometer readings are recorded. This information is used to determine the relationship between the particle size and the corresponding percentage of weight settled.

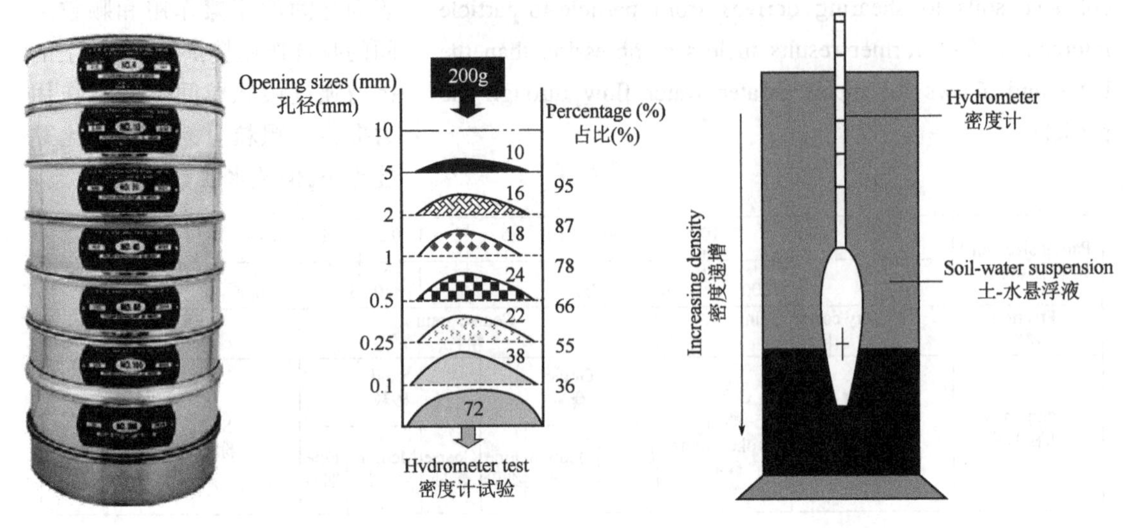

Figure 1-5 Stack of sieves

图 1-5 土工筛

Figure 1-6 Hydrometer in soil-water suspension

图 1-6 土-水悬浮液的密度计测定

结合筛分试验和密度计试验的结果,将每一级粒径与小于该粒径的土的累计质量占总质量的百分数绘制于半对数刻度表中,即为土的粒径级配曲线,如图 1-7 所示。

Data from the sieve analysis and hydrometer test are then combined. The relationship between the sieve opening size and the percentage finer than each sieve opening size are plotted on a semi-log scale to produce a grain size distribution curve, as shown in Figure 1-7.

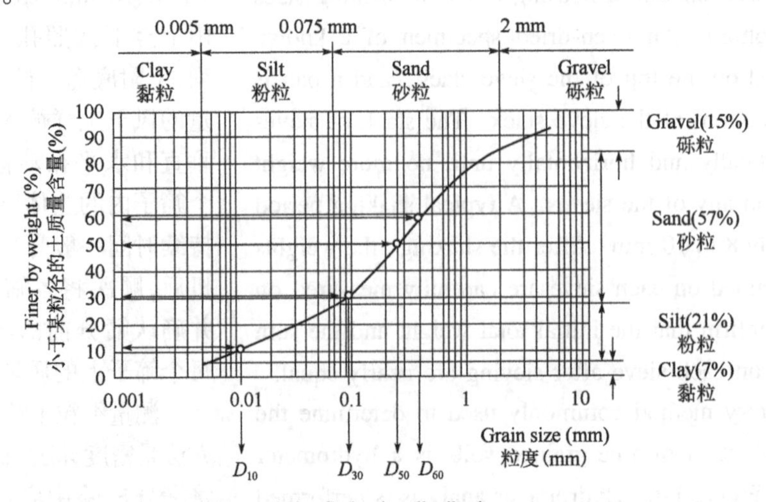

Figure 1-7 Grain size distribution curve

图 1-7 土的级配曲线

In the grain size distribution curve, multiple key grain sizes are utilized, for example, D_{10}, D_{30}, and D_{60}, which are the grain sizes corresponding to 10%, 30%, and 60% finer by weight, respectively. D_{60} is called the limiting grain size, and D_{10} is called the effective grain size. Two coefficients are defined to provide guidance on distinguishing soils based on the distribution of the particles. One of these is a numerical measure of uniformity, called the coefficient of uniformity, C_u, defined as follows:

为了进一步判断土的颗粒级配是否良好,在级配曲线上分别定出含量等于10%、30%、60%时所对应的粒径D_{10}、D_{30}和D_{60},D_{60}为控制粒径,D_{10}为有效粒径。同时,采用两个系数来反映土颗粒级配条件及曲线形态。不均匀系数C_u反映大小不同粒组的分布情况,定义为:

$$C_u = \frac{D_{60}}{D_{10}} \tag{1-1}$$

The other is the coefficient of curvature, C_c, defined as follows:

另一个系数为曲率系数C_c,定义为:

$$C_c = \frac{(D_{30})^2}{D_{10}D_{60}} \tag{1-2}$$

A soil that has $C_u \geqslant 5$ and $C_c = 1 \sim 3$ is considered a well-graded soil. A soil that is not well graded is considered uniformly graded or poorly graded. Well-graded soils achieve more stable packing because finer particles fill the voids between larger particle assemblages.

当满足$C_u \geqslant 5$且$C_c = 1 \sim 3$时,土样级配良好。当不能同时满足上述两个条件时,则为级配不良的土。级配良好的土,粗颗粒间的孔隙由细颗粒填充,土体密实,工程性质较好。

1.4 Three-phase composition of soils

1.4 土的三相组成

Soil is an assemblage of particles consisting of volumes of solids and voids[Figure 1-8a)]. The voids are occupied by air and water. Many key parameters in soil mechanics are defined by describing a soil assemblage, with solid (grain) and void (air and water) spaces, using a three-phase diagram, as shown in Figure 1-8b).

土是由固体和孔隙组成的集合体[图1-8a)],孔隙中充满了空气和水。土的很多关键参数是通过三相图中固体颗粒体积和孔隙(空气和水)体积描述的,如图1-8b)所示。

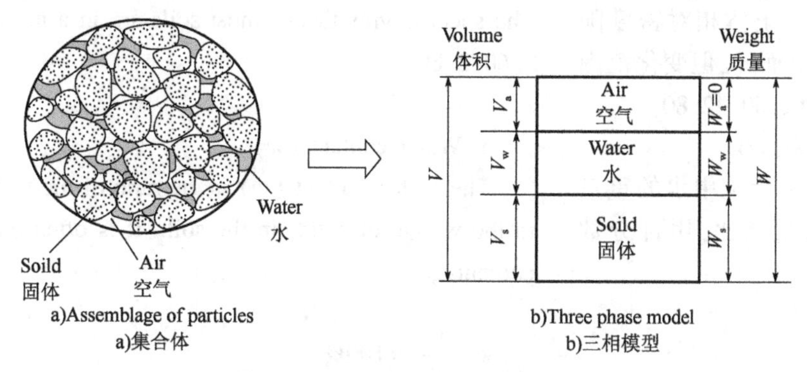

Figure 1-8 Three-phase diagram of soil

图1-8 土的三相图

用于描述土的三相组成比例关系的指标称为三相指标，包括土的密度(重度)、相对密度、含水率、孔隙比、孔隙率、饱和度、干重度、饱和重度、浮重度。

The following definitions have been established to describe the proportion of each constituent in soil, the bulk density (unit weight), specific gravity, water content, void ratio, porosity, degree of saturation, dry unit weight, saturated unit weight, and submerged unit weight.

1.4.1 实测指标

1.4.1 Measured indices

1)密度(ρ)和重度(γ)

1) Bulk density (ρ) and unit weight (γ)

密度ρ指单位体积土体的质量：

The bulk density (ρ) of a soil is the ratio of the total mass to the total volume:

$$\rho = \frac{m}{V} = \frac{m_s + m_w + m_a}{V_s + V_w + V_a} \tag{1-3}$$

密度一般用“环刀法”测定。

The bulk density is typically measured by the cutting ring method.

重度定义为单位体积土体的重量，用γ表示。

The weight of a soil per unit volume is called the bulk unit weight, denoted by γ.

$$\gamma = \frac{W}{V} = \frac{m \times g}{V} = \rho \times g \tag{1-4}$$

式中，g为重力加速度。

where g is the acceleration of gravity.

2)相对密度(G_s)

2) Specific gravity (G_s)

相对密度G_s定义为干燥土体的重量与同体积纯水重量之比：

The specific gravity (G_s) is the ratio of the weight of the soil solids to the weight of an equal volume of water, given as follows:

$$G_s = \frac{W_s}{V_s \gamma_w} = \frac{\gamma_s}{\gamma_w} \tag{1-5}$$

式中，γ_w为水的重度，取9.81 kN/m^3。

where γ_w is the unit weight of water, 9.81 kN/m^3.

相对密度在实验室中用比重瓶法测定。土体相对密度随土粒矿物成分而异，但变化范围不大，通常为2.60~2.80。

A pycnometer is used to determine the specific gravity. The specific gravities of most soils are in a narrow range of 2.60-2.80.

3)含水率(w)

3) Water content (w)

含水率w为土中水的重量与土颗粒重量之比，用百分数表示：

The water content (w) is the ratio of the weight of water to the weight of solids in the soil. It is often expressed as a percentage:

$$w = \frac{W_w}{W_s} \times 100\% \tag{1-6}$$

The weight of water is determined by weighing a sample of the soil and then placing it in an oven at (110 ± 5)℃ until the weight of the sample remains constant-that is, until all the absorbed water evaporates.

土中水质量的测定方法是：取原状土样称取质量，然后将土样在(110 ± 5)℃环境温度烘箱内烘至恒量，再称取烘干后土的质量，原状土与干土的质量之差即为土中水质量。

1.4.2 Deduced indices

1.4.2 推导指标

1) Void ratio (e)

1)孔隙比(e)

The void ratio (e) is the ratio of the volume of void space to the volume of solids and is usually expressed as a decimal quantity:

孔隙比 e 为土中孔隙体积与土颗粒体积之比，孔隙比通常采用小数表示：

$$e = \frac{V_v}{V_s} \tag{1-7}$$

2) Porosity (n)

2)孔隙率(n)

The porosity (n) is the ratio of the volume of voids to the total volume and is usually expressed as a percentage:

孔隙率 n 表示孔隙体积占总体积的百分数：

$$n = \frac{V_v}{V} \tag{1-8}$$

3) Degree of saturation (S)

3)饱和度(S)

The degree of saturation (S) is the ratio of the volume of water to the volume of voids. It is often expressed as a percentage:

饱和度 S 指土中孔隙被水充满的程度，即孔隙中水所占体积与孔隙总体积之比：

$$S = \frac{V_w}{V_v} \tag{1-9}$$

The value of S ranges from 0% for completely dry soil to 100% for fully saturated soil. Soils with degree of saturation values in the range of $0 < S < 100\%$ are called partially saturated soils.

$S = 0\%$，表示土体完全干燥；$S = 100\%$，表示土体完全饱和；$0\% < S < 100\%$ 表示土体部分饱和。

4) Dry unit weight (γ_d)

4)干重度(γ_d)

The dry unit weight γ_d is the weight of the soil particles ($S = 0\%$) in a unit volume:

干重度表示单位体积的干土($S = 0\%$)重量：

$$\gamma_d = \frac{W_s}{V} \tag{1-10}$$

The dry unit weight depends on the void ratio of the soil. A lower void ratio corresponds to a denser soil. The dry unit weight is often used as an indicator of the compaction quality of soil fill.

干重度取决于土体中孔隙的多少，e 越小，γ_d 越大，即土体越密实。因此，工程上常以干重度作为衡量土体压实状态的指标。

5)饱和重度(γ_{sat})

饱和重度指土体孔隙中完全充满水($S=100\%$)时的重度:

5) Saturated unit weight (γ_{sat})

The weight of a unit volume of a saturated soil ($S=100\%$) is defined as follows:

$$\gamma_{sat}=\frac{W_s+V_v\gamma_w}{V} \tag{1-11}$$

6)浮重度(γ')

浮重度γ'指土体在水下的单位重量:

6) Buoyant unit weight (γ')

The buoyant unit weight, γ', of a soil refers to its unit weight under water:

$$\gamma'=\frac{W_s-V_s\gamma_w}{V} \tag{1-12}$$

1.4.3 换算关系

换算关系的计算模型如图1-9所示。

1.4.3 Conversion relationships

The calculation model of the conversion relationship is shown in Figure 1-9.

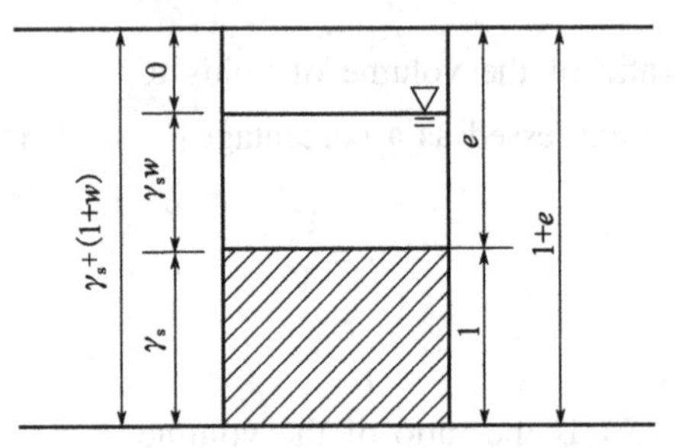

Figure 1-9 Computing model (assume $\gamma_s=1$)

图1-9 计算模型(假设$\gamma_s=1$)

假定$V_s=1$,则$W_s=V_s\cdot\gamma_s$,由于$w=W_w/W_s$,因此$W_w=w\cdot\gamma_s$。

由于$W_a=0$,因此$W=W_s+W_w=\gamma_s\cdot(1+w)$。

由于$e=V_v/V_s$,因此$V_v=e$,$V=1+e$。

$\gamma=W/V$,因此$\gamma=\gamma_s\cdot(1+w)/(1+e)$,可得:

Assuming $V_s=1$, $W_s=V_s\cdot\gamma_s$. $w=W_w/W_s$, hence, $W_w=w\cdot\gamma_s$.

$W_a=0$, hence, $W=W_s+W_w=\gamma_s\cdot(1+w)$.

$e=V_v/V_s$, hence, $V_v=e$ and $V=1+e$.

$\gamma=W/V$, hence, $\gamma=\gamma_s\cdot(1+w)/(1+e)$. Then:

$$e=\frac{\gamma_s(1+w)}{\gamma}-1 \tag{1-13}$$

由式(1-13)可知,孔隙比e可以由3个实测指标计算获取,这3个实测指标可在实验室测定。其他指标可以直接采用e来表示,可见e是所有导出指标的“桥梁”。

The void ratio e is the “bridge” of all the derived indices, because Equation (1-13) shows that e can be expressed by three measured indices, which can be measured from laboratory tests; thus, the void ratio e can be determined, and other indices can be expressed directly in terms of e.

We have the following relationships:

包括以下关系：

$$n = \frac{e}{1+e} \tag{1-14}$$

$$S = \frac{V_w}{V_v} = \frac{\gamma_s w}{e\gamma_w} \tag{1-15}$$

$$\gamma_d = \frac{W_s}{V} = \frac{\gamma_s}{1+e} = \frac{\gamma}{1+w} \tag{1-16}$$

$$\gamma_{sat} = \frac{\gamma_s + e\gamma_w}{1+e} \tag{1-17}$$

$$\gamma' = \frac{\gamma_s - \gamma_w}{1+e} = \gamma_{sat} - \gamma_w \tag{1-18}$$

Example 1-1

For a given soil, values of $w = 25\%$ and $\gamma = 18.5\ \text{kN/m}^3$ are measured. Determine the void ratio e and the degree of saturation S. Assume that G_s is 2.70.

例题 1-1

对于给定的土样，测得 $w = 25\%$ 和 $\gamma = 18.5\ \text{kN/m}^3$。请计算孔隙比 e 和饱和度 S。假设 G_s是2.70。

Solution

First, assume $W_s = 100$ kN; then $W_w = 100 \times 0.25 = 25$ kN.

解：

首先，假设 $W_s = 100$ kN；那么 $W_w = 100 \times 0.25 = 25$ kN。

$$V_s = W_s/\gamma_s = W_s/(G_s\gamma_w) = 100/(2.7 \times 9.81) = 3.775(\text{m}^3)$$

$$V_w = W_w/\gamma_w = 25/9.81 = 2.548(\text{m}^3)$$

Calculate V_a from γ:

由 γ 反算 V_a：

$$\gamma = 18.5\ \text{kN/m}^3 = (W_s + W_a)/(V_s + V_w + V_a) = (100 + 25)/(3.775 + 2.548 + V_a)$$

$$V_a = 0.434(\text{m}^3)$$

With all components in the three phases obtained, then:

在计算出所有三相分量的情况下可得：

$$e = \frac{V_w + V_a}{V_s} = \frac{2.548 + 0.434}{3.775} = 0.790$$

$$S = \frac{V_w}{V_w + V_a} = \frac{2.548}{2.548 + 0.434} = 85.4\%$$

Example 1-2

In its natural condition, a soil sample has a mass of 2290 g and a volume of 1.15×10^{-3} m^3. After being completely dried in an oven, the mass of the sample is 2035 g. The value of G_s for

例题 1-2

某原状土样质量为2290 g，体积为 1.15×10^{-3} m^3。在烘箱中完全烘干后，土样质量为2035 g，土

体 G_s 值为2.68。请计算含水率、密度、重度、孔隙比、孔隙率和饱和度。

the soil is 2.68. Determine the water content, bulk density, unit weight, void ratio, porosity, and degree of saturation.

解:

根据式(1-3)、式(1-4)和式(1-6),可以直接计算土体的堆积密度 ρ、重度 γ 和含水率 w。

Solution

According to Equations (1-3), (1-4), and (1-6), the bulk density ρ, unit weight γ, and water content w can be calculated directly by definition.

$$\rho = \frac{m}{V} = \frac{2.29}{1.15 \times 10^{-3}} = 1990(\mathrm{kg/m^{-3}})$$

$$\gamma = \frac{mg}{V} = \rho g = 1990 \times 9.81 = 19.5(\mathrm{kN/m^3})$$

$$w = \frac{W_w}{W_s} = \frac{2290 - 2035}{2035} \times 100\% = 12.5\%$$

孔隙率 e、孔隙比 n 和饱和度 S 通过式(1-13)、式(1-14)和式(1-15)计算确定。

The void ratio, porosity, and degree of saturation can be determined from phase relationships using Equations (1-13), (1-14), and (1-15).

$$e = \frac{\gamma_s(1+w)}{\gamma} - 1 = \frac{G_s\gamma_w(1+w)}{\gamma} - 1 = \frac{2.68 \times 9.81 \times (1+0.125)}{19.5} - 1 = 0.52$$

$$n = \frac{e}{1+e} = \frac{0.52}{1.52} \times 100\% = 34\%$$

$$S = \frac{wG_s}{e} = \frac{2.68 \times 12.5\%}{0.52} = 64.5\%$$

1.5 土的物理状态

1.5 Physical states of soils

1.5.1 细粒土的稠度界限值和指标

1.5.1 Atterberg limits and indices of fine-grained soils

细粒土的物理和力学特性随着含水率的增大分别呈现固态、半固态、可塑状态和流动状态4种不同的状态。当黏土颗粒中含有大量的水时,土体呈液态;当在稍微干燥状态下,呈可塑状态,类似于软黄油;当更为干燥时,呈半固态,类似于奶酪;当处于非常干燥的阶段时,呈固态,类似于硬糖,如图1-10所示。

The physical and mechanical properties of fine-grained soils are linked to four distinct states: solid, semi-solid, plastic, and liquid, in the order of increasing water content. A mixture of clay particles with a large amount of water is in a liquid state. In a slightly drier (plastic) state, it becomes similar to soft butter. When it is dried further, it reaches a semi-solid state, similar to cheese. At a very dry (solid) stage, it is similar to a hard candy. These stages are illustrated in Figure 1-10.

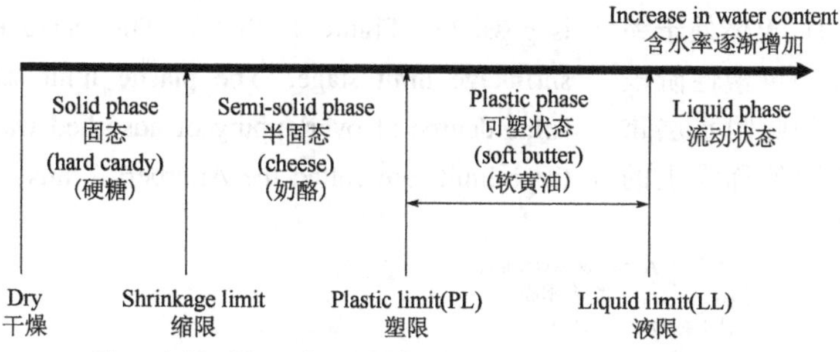

Figure 1-10 Phase change of fine-grained soil with water content

图 1-10 黏性土相态随含水率的变化

The shrinkage limit (w_S) is the maximum water content below which no further volume change of the soil occurs, as illustrated in Figure 1-11. Note that, at the shrinkage limit, the soil may be still fully saturated. The plastic limit (w_P) is the boundary water content between the plastic and semi-solid phases; and the liquid limit (w_L) of a soil-water mixture is the boundary water content between the liquid and plastic phases.

缩限(w_S)指由半固态变为固态的界限含水率。低于该含水率,黏土不会发生体积变化,如图 1-11 所示。注意,此时土体仍然可能是完全饱和的。塑限(w_P)指由半固态变为可塑状态的界限含水率。液限(w_L)指由可塑状态变为流动状态的界限含水率。

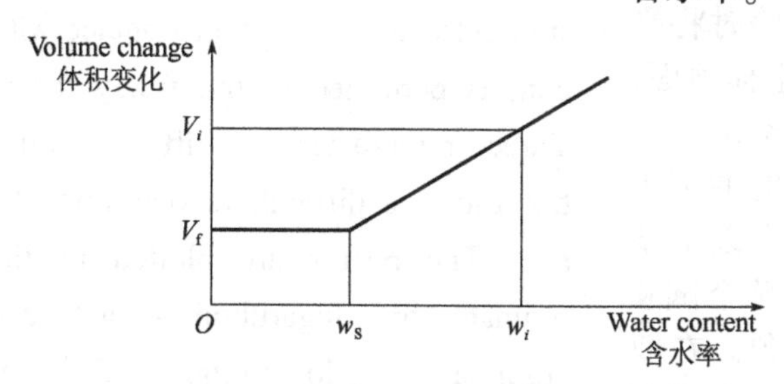

Figure 1-11 Definition of shrinkage limit

图 1-11 缩限的定义

The liquid limit w_L and shrinkage limit w_S can be determined based on the knowledge of the soil-water system. The adsorbed water layer is considered an integral part of a fine-grained soil particle. As shown in Figure 1-12, when clay particles contain enough water, adsorbed water layers are not at all in contact with each other, and thus there is no frictional resistance. This is a liquid stage [Figure 1-12a)]. If water is removed to a certain level at which all of the adsorbed water layers come into contact, frictional resistance is developed at the contact points. This is considered to be the liquid limit stage [Figure 1-12b)]. If the soil-water mixture is dried further, the adsorbed water layers overlap. The limiting stage of this overlapping is the level at which the particles themselves touch each other and no further overlapping

液限 w_L 和缩限 w_S 可以通过土中水分形态的变化来确定。结合水膜层是细粒土的土颗粒的重要组成部分。如图 1-12 所示,当黏土颗粒含有大量水时,结合水膜互不接触,土颗粒之间不存在摩擦阻力,土体处于流动状态[图 1-12a)];当黏土含水率减少至土颗粒表面结合水膜处于相互接触时,土颗粒接触点之间就会产生摩擦阻力,土体处于液限阶段[图 1-12b)];当土体进一步干燥时,结合水膜发生重叠,当土颗粒表面结合水膜到达极限而无法继续重叠时,土体处于缩限阶段[图1-12c)]。塑性阶段可能有一定程度的吸附水层重叠。上述三个界限值称为土的稠度界限含水率。

is possible [Figure 1-12c)]. This stage is considered the shrinkage limit stage. The plastic limit stage may involve some degree of overlapping of adsorbed water layers. These three limits are called the Atterberg limits.

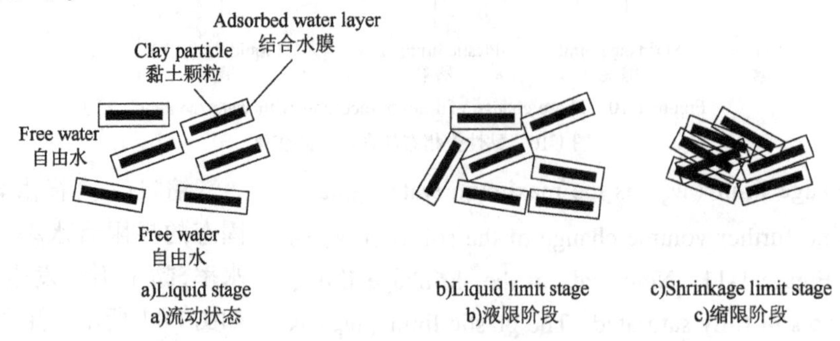

Figure 1-12 Fine-grained soil particles with adsorbed water layers in water

图1-12 黏土颗粒的结合水

液限和塑限采用锥式液限仪进行测定。如图1-13所示,落锥试验中的圆锥仪锥体尖角为30°,质量为76 g。试验时将平衡锥放在已制备好的土样顶面,使其在重力作用下徐徐沉入试样,测定圆锥仪在5 s时的下沉深度。一般需要做3个以上不同含水率的试样,因为单个试验很难得到准确的界限值。在双对数坐标上绘制圆锥下沉深度和含水率的试验结果,并拟合为直线,如图1-14所示。圆锥下沉深度为17 mm时所对应的含水率为液限,下沉深度为2 mm时所对应的含水率为塑限。

A fall cone test is an accurate method of determining both the liquid and plastic limit. In a fall cone test (Figure 1-13), a cone with an apex angle of 30° and a total mass of 76 g is suspended above, yet in contact with, a soil sample. The cone is permitted to fall freely for a period of 5 seconds. Three or more tests at different water contents are required because it is difficult to determine the limits from a single test. The results are plotted as the water content (the ordinate, on a logarithmic scale) versus the penetration (the abscissa, on a logarithmic scale), and the best-fit straight line (the liquid state line) linking the data points is drawn (Figure 1-14). Based on the plot, the liquid limit (w_L) is the water content on the liquid state line corresponding to a penetration of 17 mm. The plastic limit (w_P) is the water content at a penetration of 2 mm.

塑性指数(I_P)是表征土体处于塑性状态时含水率范围的指标,定义为:

The plasticity index (I_P) is a measure of the water content range over which a material exhibits plastic behavior. It is defined as:

$$I_P = w_L - w_P \tag{1-19}$$

黏性土的许多工程力学特性与塑性指数相关。

Many engineering properties of clays are related to I_P.

液性指数(I_L)指土的天然或原位含水率与其塑限含水率之间的差值与塑性指数的比值,定义为:

Another parameter, the liquidity index (I_L), is defined as the ratio of the difference between the natural or in situ water content of a soil and its plastic limit to its plasticity index. It is expressed as:

$$I_L = \frac{w_n - w_P}{I_P} \tag{1-20}$$

Where w_n is the natural water content of the soil.

式中,w_n为土体的天然含水率。

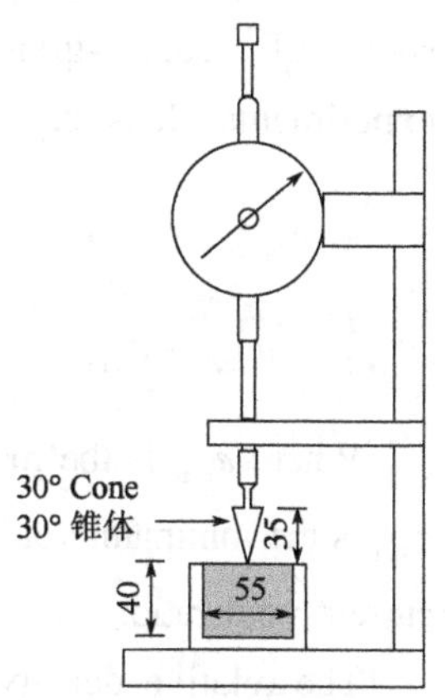

Figure 1-13 Fall cone apparatus (unit: mm)

图 1-13 锥式液限仪(单位:mm)

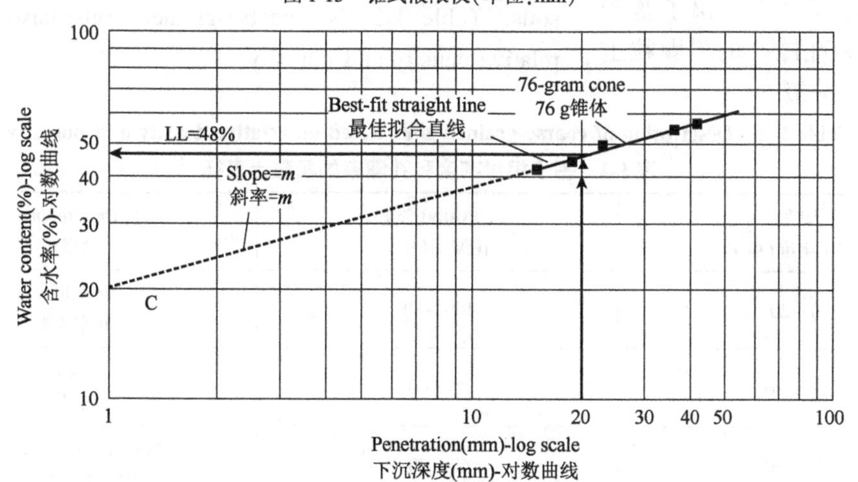

Figure 1-14 Typical fall cone test results

图 1-14 典型的圆锥仪试验结果

Table 1-1 summarizes the relationship between soil strength and I_L.

表 1-1 归纳了土体强度和 I_L 之间的关系。

Table 1-1 Relationship between liquidity index and the strength of fine-grained soils

表 1-1 基于液性指数的细粒土强度特性

Values of I_L 液性指数 I_L	Description of soil strength 土体强度表述
$I_L \leqslant 0$	①Semi-solid state-high strength, brittle, (sudden) fracture is espected ①半固态,高强度,脆性,容易出现断裂
$0 < I_L \leqslant 1$	②Plastic state-intermediate strength, soil deforms like a plastic material ②塑性状态,中等强度,土体变形类似塑性材料
$I_L > 1$	③Liquid state-low strength, soil deforms like a viscous fluid ③液态,强度低,土体变形类似黏性流体

1.5.2 粗粒土的相对密度 D_r

相对密度 D_r 指粗粒土的天然孔隙比 e 与其在最松散状态下的孔隙比 e_{max} 和在最密实状态下的孔隙比 e_{min} 之间的关系。e_{max} 和 e_{min} 可通过试验确定。其表示为:

1.5.2 Relative density of coarse-grained soils D_r

Relative density (D_r) is an index that quantifies the degree of packing between the loosest and densest possible state of coarse-grained soils, as determined from experiments. It is expressed as:

$$D_r = \frac{e_{max} - e}{e_{max} - e_{min}} \tag{1-21}$$

式中,e_{max} 为最大孔隙比(最松散状态);e_{min} 为最小孔隙比(最密实状态);e 为当前孔隙比。

相对密度与粗粒土的强度密切相关,密实土体的强度优于松散土体。表 1-2 列出了基于相对密度和孔隙率的粗颗粒土的密实程度分级。

Where e_{max} is the maximum void ratio (loosest condition), e_{min} is the minimum void ratio (densest condition), and e is the current void ratio.

The relative density correlates well with the strength of coarse-grained soils, denser soils being stronger than looser soils. Table 1-2 lists coarse-grained soils based on their relative density and porosity.

Table 1-2 Description of coarse-grained soils based on relative density and porosity

表 1-2 基于相对密度和孔隙率的粗粒土表征

D_r(%) 相对密度(%)	Porosity(%) 孔隙率(%)	Description 类型
0 ~ 20	100 ~ 80	Very loose 非常松散
20 ~ 40	80 ~ 60	Loose 松散
40 ~ 70	60 ~ 30	Medium dense or firm 中等密实或坚硬
70 ~ 85	30 ~ 15	Dense 密实
85 ~ 100	<15	Very dense 非常密实

1.6 土的工程分类

由于起源、成分、位置、地质历史和许多其他因素的不同,自然界存在着种类繁多、性质各异的土。原位试验和室内试验是获取土体各类指标参数值和工程特性的重要途径。为了工程实践的需要,将土体按照其工程

1.6 Engineering classification of soils

Soils differ depending on their origins, compositions, locations, geological histories, and many other factors. In situ and laboratory tests on soil specimens are critically important to determine their index parameter values and engineering characteristics. However, it is convenient for engineers to categorize soils into groups with similar engineering behaviors. This process is called soil classification.

Most soil classification standards use soil indices, such as Atterberg limits (liquid limit and plastic limit) and soil gradation information (D_{10}, D_{60}, C_u, and C_c). Several standards are widely used in current geotechnical engineering practice in China, including the following:

Standard for Engineering Classification of Soil (GB/T 50145—2007)

Code for Design of Building Foundations (GB 50007—2011)

Test Methods for Soils for Highway Engineering (JTG 3430—2020)

The soil classifications for inorganic soil based on grain-size characteristics are huge-grained, coarse-grained, and fine-grained soils.

1) Very coarse-grained soils

Very coarse-grained soils are soils whose coarse particles (size >2 mm) make up more than 50% of the total weight. The particle sizes and gradations of these types of soil greatly influence their behavior in engineering construction. These types are classified as shown in Table 1-3.

特性进行分组,称为土的分类。

土的分类标准大多以稠度界限含水率(液限、塑限)与颗粒级配(D_{10}、D_{60}、C_u和C_c)等指标为依据。国内岩土工程通常采用以下分类标准:

《土的工程分类标准》(GB/T 50145—2007)

《建筑地基基础设计规范》(GB 50007—2011)

《公路土工试验规程》(JTG 3430—2020)

根据粒径特征,无机土分为巨粒土、粗粒土和细粒土。

1)碎石土

碎石土指粒径大于2 mm颗粒含量超过土体总重50%的土。该类土的粒径和级配对工程建设影响显著。根据粒组含量及颗粒形状,碎石土的分类见表1-3。

Table 1-3 Classification of Very coarse-grained soils

表1-3 巨粒土的分类

Soil type 土的名称	Soil particle shape 颗粒形状	Gradation 粒组含量
Boulder 漂石	Round forms predominant 圆形为主	Particles with sizes >200 mm contribute to more than 50% of the total weight 粒径大于200 mm的颗粒占总重量的50%以上
Block stone 块石	Angular forms predominant 棱角形为主	
Cobble 卵石	Round forms predominant 圆形为主	Particles with sizes >20 mm contribute to more than 50% of the total weight 粒径大于20 mm的颗粒占总重量的50%以上
Broken stone 碎石	Angular forms predominant 棱角形为主	
Round gravel 圆砾	Round forms predominant 圆形为主	Particles with sizes >2 mm contribute to more than 50% of the total weight 粒径大于2 mm的颗粒占总重量的50%以上
Angular gravel 角砾	Angular forms predominant 棱角形为主	

2)砂土

砂土指粒径大于2 mm的颗粒含量不超过土体总重的50%,而粒径大于0.075 mm的颗粒含量超过总重50%的土。砂土的分类标准类似于碎石土,见表1-4。

2) Sand

Sands are soils in which particles with sizes > 2 mm contribute to less than 50% of the total weight and particles > 0.075 mm contribute to more than 50% of the total weight. Sands are classified in a manner similar to crushed stones and gravels, as shown in Table 1-4.

Table 1-4 Classification of sand based on gradation

表1-4 基于粒组含量的砂土分类

Soil type 土的名称	Gradation 粒组含量
Gravelly sand 砾砂	Particles >2 mm contribute to 25% ~50% of the total weight 粒径大于2 mm的颗粒占总重量的25% ~50%
Coarse sand 粗砂	Particles >0.5 mm contribute to 50% of the total weight 粒径大于0.5 mm的颗粒占总重量的50%
Medium sand 中砂	Particles >0.25 mm contribute to 50% of the total weight 粒径大于0.25 mm的颗粒占总重量的50%
Fine sand 细砂	Particles >0.075 mm contribute to 85% of the total weight 粒径大于0.075 mm的颗粒占总重量的85%
Silty sand 粉砂	Particles >0.075 mm contribute to 50% of the total weight 粒径大于0.075 mm的颗粒占总重量的50%

3)粉土

粉土指粒径大于0.075 mm的颗粒含量小于总重50%且塑性指数$I_P \leqslant 10$的土。

3) Silts

Silts are soils with plasticity index values $I_P \leqslant 10$ in which particles > 0.075 mm make contribute to less than 50% of the total weight.

4)黏性土

根据《建筑地基基础设计规范》(GB 50007—2011)和《铁路工程岩土分类标准》(TB 10077—2019),黏性土依据塑性指数进行分类。$10 < I_P \leqslant 17$的称为粉质黏土;$I_P > 17$的称为黏土。

4) Clay

According to the *Code for Design of Building Foundations* (GB 50007—2011) and the *Code for Rock and Soil Classification for Railway Engineering* (TB 10077—2019), cohesive soils are classified based on their plasticity index I_P. $10 < I_P \leqslant 17$, Silty clay; $I_P > 17$, Clay

习　题

(1)某土样$G_s = 2.72$,$e = 0.95$,$S_r = 37\%$,如要将S_r提高到90%,则每立方米土样应该加多少水?

Exercises

(1) For a certain soil sample, $G_s = 2.72$, $e = 0.95$, and $S_r = 37\%$. How much water should be added per cubic meter of the soil sample if S_r is to be increased to 90%?

(2) The density of a dry sand sample is 1.66 g/cm^3, and the specific gravity of the soil particles is 2.70. If the dry sand sample is put in the rain and the volume of the sand sample remains unchanged while the saturation increases to 0.60, what are the density and water content of the wet sand?

(2) 某干砂试样的密度为 1.66 g/cm^3,相对密度为 2.70,将此干砂试样置于雨中,若砂样体积不变,饱和度增加到 0.60,计算湿砂的密度和含水率。

(3) The saturated unit weight (γ_{sat}) of the sand layer below the groundwater level is 19.91 kN/m^3; the relative density D_r is 2.66; the maximum dry weight is 16.7 kN/m^3; and the minimum dry weight is 13.9 kN/m^3. Judge the compactness of the sand.

(3)某地下水位以下砂层的饱和重度 γ_{sat} = 19.91 kN/m^3,相对密度 D_r = 2.66,最大干重度 16.7 kN/m^3,最小干重度 13.9 kN/m^3,试判断该砂土的密实程度。

(4) The weight of an undisturbed soil sample is 1.84 kN; the weight of the dry soil is 1.44 kN; the plastic limit is 5%; and the liquid limit is 27%. Calculate the density and water content of the wet sand.

(4) 某原状土重量为 1.84 kN,干土重量为 1.44 kN,塑限为 5%,液限为 27%,试确定该土样的名称及状态。

(5) The water content of a sand is 28.5%; the unit weight of the soil is 19 kN/m^3, and the specific gravity of the soil is 2.68. The particle analysis results are shown in the Table 1-5.

(5) 某砂土的含水率 w = 28.5%,土的天然重度 γ = 19 kN/m^3,土粒相对密度 D_r = 2.68,颗粒分析结果见表 1-5。

Table 1-5 The particle analysis results of Exercises (5)

表 1-5 习题(5)颗粒级配分析结果

Particle size range of soil particle group(mm) 土粒组的粒径范围(mm)	>2	2 ~ 0.5	0.5 ~ 0.25	0.25 ~ 0.075	<0.075
Percentage of grain fraction in total dry soil mass(%) 粒组占干土总质量的百分数(%)	9.4	18.6	21.0	37.5	13.5

Requirement:

①Determine the type of this soil sample;

②Calculate the void ratio and degree of saturation of the soil.

要求:

①确定该土样的种类;

②计算该土样的孔隙比和饱和度。

Chapter 2 Stress Distribution in Soils 土中应力分布

2.1 半无限体与侧限应力状态

将天然地面视为一个平面，假定地基土为半无限体，如图2-1所示，以水平地面为界，在 X、Y 轴的正负方向和 Z 轴的正方向与建筑物的尺寸相比都可以认为是无限的，故称为半无限体。

2.1 Semi-infinite body and confined stress state

The natural ground is regarded as a plane, and a foundation soil is assumed to be a semi-infinite body. As shown in Figure 2-1, taking the horizontal ground as the boundary, the positive and negative directions of the X axis and Y axis and the positive direction of the Z axis can be considered as infinite in comparison to the size of buildings; therefore, a body of soil bounded by the natural ground surface is called semi-infinite body.

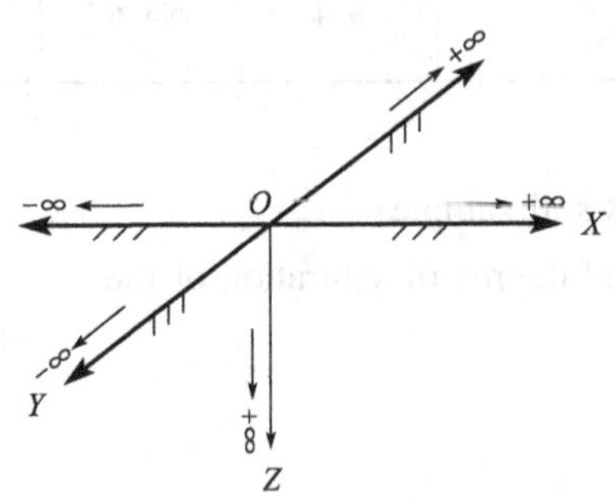

Figure 2-1 Semi-infinite body

图2-1 半无限体

当土质均匀时，则任一水平面上的竖向自重应力都是均匀无限分布的。在此应力作用下，地基土只能产生竖向变形，不能产生侧向变形。土体内任一竖直面都是对称面，对称面上的剪应力为零。根据剪应力互等定理可知，任一水平面上的剪应力

When the soil is uniform, the vertical self weight stress on any horizontal plane is uniformly and infinitely distributed. Under the action of this stress, the foundation soil can only exhibit vertical deformation; it is impossible for it to exhibit lateral deformation. Any vertical surface in the soil is a symmetrical plane, and the shear stress on the symmetrical plane is zero. According to the reciprocal theorem of shear stress, the shear stress on any horizontal plane is also equal to zero. If a soil column with an area of A is cut from the soil, as shown in Figure 2-2, according to the static equilibrium conditions, the vertical self weight stress of the soil column at depth Z is equal to the weight of the soil column per unit area.

也等于零。若在土中切取一个面积为 A 的土柱,如图 2-2 所示,根据静力平衡条件可知,在 Z 深度处的平面,土柱自重产生的竖向自重应力等于单位面积土柱的重量。

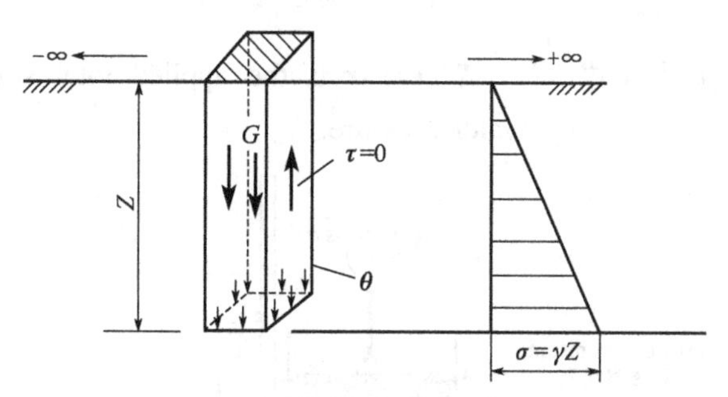

Figure 2-2 Stresses on soil column

图 2-2 土柱受力情况

2.2 Principle of effective stress

Soil is made up of particles assembled; thus, the soil's skeleton (the particle-connected structure) is a body that resists external forces, as shown in Figure 2-3. Two-headed arrow vectors indicate interparticle forces at contact points, including normal and shear contact forces. In a dry situation, the interparticle forces are in equilibrium with the external forces. However, if the soil is saturated or partially saturated, pore water pressure develops; therefore, the pore water resists some portion of the external forces.

2.2 有效应力原理

土是由土颗粒组成的,因此土骨架是抵抗外部荷载的主体,如图 2-3 所示。双向箭头矢量表示土粒接触点处的作用力,包括法向接触力和切向接触力。土体干燥时,土粒间接触力和外力处于平衡状态。但当土体饱和或部分饱和时,会存在孔隙水压力抵抗部分外力。

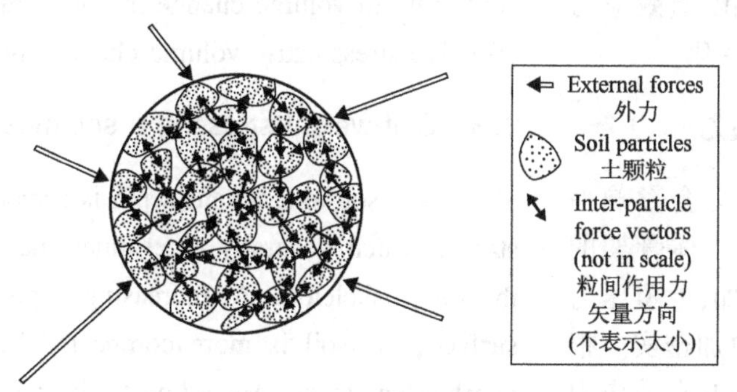

Figure 2-3 Interparticle stresses in particle assemblage

图 2-3 土粒间接触应力

Figure 2-4 illustrates a model of the interparticle and pore water pressure components of resistance to the external stress σ. The model consists of a water-filled cylinder with a frictionless

图 2-4 为土颗粒和孔隙水压力抵抗外部应力 σ 的模型。该模型由一个充水圆筒和一个

弹簧支撑的活塞组成。弹簧代表土骨架的阻力,水代表土体中的孔隙水。活塞上有一个小孔,可以排水,代表土的渗透性。太沙基将有效应力 σ' 定义为:

piston that is supported by a spring. The piston has a small hole to allow drainage. The spring represents the skeleton's resistance; the filled water represents the pore water in the soil, and a small hole in the piston reflects the permeability of the soil. Terzaghi defined the effective stress σ' as:

$$\sigma' = \sigma - u \tag{2-1}$$

式中,σ 为总应力;u 为孔隙水压力。

Where σ is the applied total stress and u is the pore water pressure.

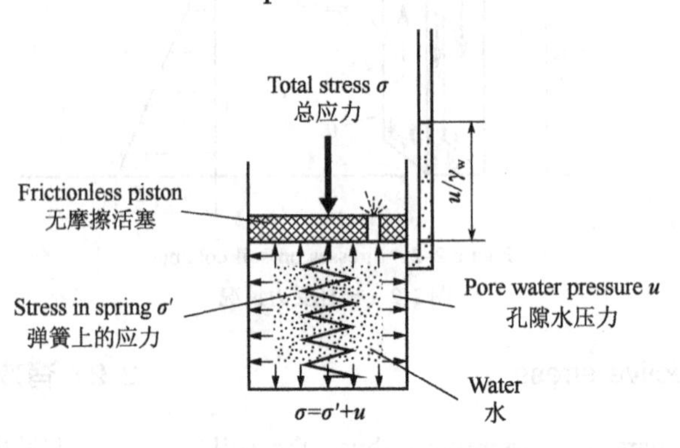

Figure 2-4 Terzaghi's effective stress model

图 2-4 太沙基有效应力模型

在模型中,外部应力一部分由弹簧应力 σ' 承担,另一部分由孔隙水压力 u 承担,因此土骨架应力与有效应力密切相关。当体积发生变化,即弹簧被压缩时产生有效应力 σ',反之亦然。总应力不影响土的体积,有效应力才会引起土的体积变化。

In the model, the applied stress is carried partially by stress in spring σ' and partially by the pore water pressure u. Thus, the skeleton's stress is closely related to the effective stress. When a volume change occurs (the spring is compressed), σ' (effective stress) develops, or vice versa, as demonstrated by the model. The total stress does not contribute to volume change in soils; rather, changes in the effective stress cause volume changes in soils.

2.3 土体的自重应力

2.3 Self-weight stress in a soil mass

上覆土体的重力会影响一定深度位置的原位土体骨架的形成。随着埋深增加,上覆压力逐渐增大,土体会愈加密实。根据有效应力原理可知,有效应力决定土骨架形态。下面介绍各种情况下土体有效自重应力的计算方法。

In situ soil at a certain depth is subjected to an overburden stress, which generally determines the current formation of the soil's skeleton. At increasing depth beneath the ground surface, the soil is more compacted because of the higher overburden stress. According to the effective stress concept, the stress that determines the current form of the skeleton is the effective stress. In the following text, effective overburden stress computations are demonstrated for various situations.

Figure 2-5 shows several layers of dry soil deposit. The total vertical (overburden) stress at Point A is the weight of a soil column with a unit area above Point A. Thus:

图 2-5 为数层干燥土体沉积物。作用在 A 点的总竖向应力为上部各个土层的单位截面积土柱重量的叠加，因此：

$$\sigma = H_1\gamma_1 + H_2\gamma_2 + H_3\gamma_3 = \sum_{i=1}^{3}(H_i\gamma_i) \tag{2-2}$$

The vertical stress distribution σ with depth is plotted alongside the soil columns. In this case, $u = 0$, and thus, $\sigma' = \sigma$ throughout the depth.

总竖向应力 σ 随深度的分布特征如图 2-5 所示，干土层的孔隙水压力 $u = 0$，因此 $\sigma' = \sigma$。

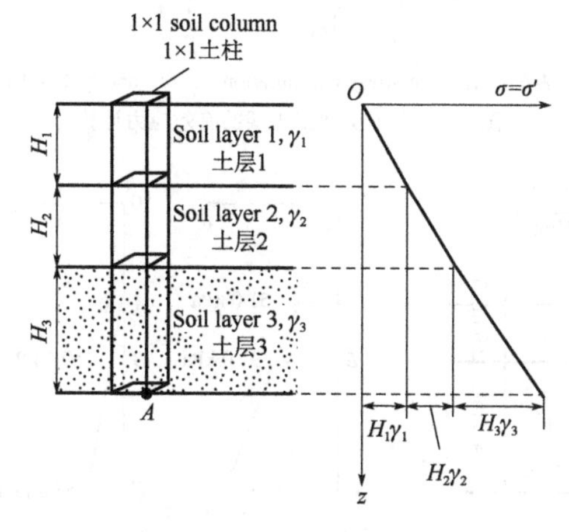

Figure 2-5 Effective stress computation for dry soil layers

图 2-5 干土层的有效应力计算

Figure 2-6 illustrates a situation with a steady groundwater table. The water table is at the mid-depth of soil layer 2. The total vertical stress σ at Point A is calculated first as the weight of a soil column with unit area, as before, and then the hydrostatic water pressure u is computed. Finally, the effective vertical stress σ' is computed as follows:

图 2-6 为一个具有稳定地下水位的工况，地下水位处于土层 2 的中间。首先计算 A 点的总竖向应力 σ，然后计算孔隙水压力 u。那么，有效自重应力 σ' 的计算结果如下：

$$\sigma = H_1\gamma_1 + H_2\gamma_2 + H_3\gamma_3 + H_4\gamma_4 = \sum_{i=4} H_i\gamma_i \tag{2-3}$$

$$u = (H_3 + H_4)\gamma_w \tag{2-4}$$

$$\begin{aligned}\sigma' &= \sigma - u = [H_1\gamma_1 + H_2\gamma_2 + H_3\gamma_3 + H_4\gamma_4] - [(H_3 + H_4)\gamma_w] \\ &= \sum_{i=1,2}(H_i\gamma_i) + \sum_{j=3,4}[H_j(\gamma_j - \gamma_w)]\end{aligned} \tag{2-5}$$

Example 2-1

A foundation is composed of multi-layer soils, as shown in the geological section in Figure 2-7a). Calculate and draw the total normal stress distribution, pore water pressure distribution, and effective normal stress distribution of the soils in the depth direction.

例题 2-1

一个地基由多层土体组成，地质剖面如图 2-7a) 所示。计算并绘制土体沿深度方向的总竖向应力分布、孔隙水压力分布和有效自重应力分布。

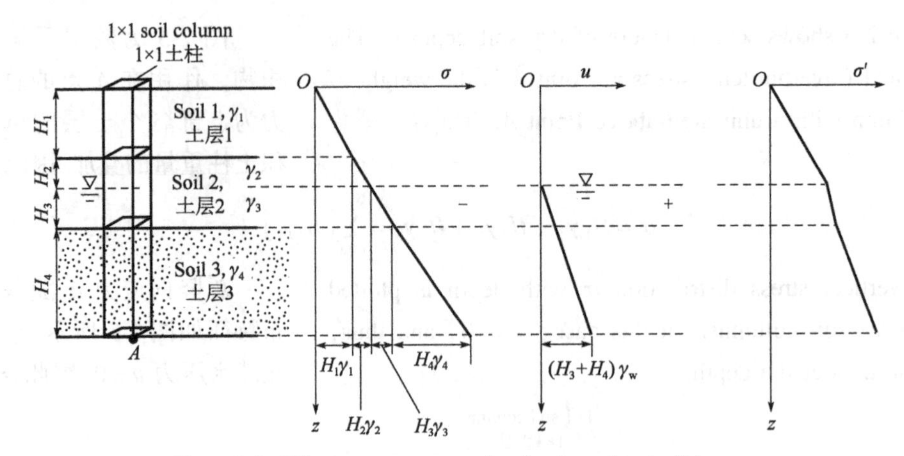

Figure 2-6 Effective stress computation for dry and wet soil layers

图 2-6 干土层和湿土层的有效应力计算

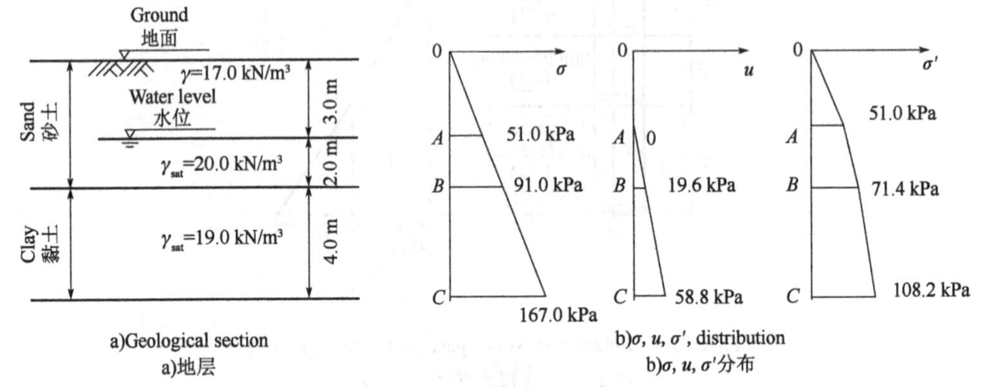

Figure 2-7 Diagram for example 2-1

图 2-7 例题 2-1 的附图

解:

取 A、B、C 三个点作为参考点，σ，u，σ' 的计算结果如表 2-1 所示，数值标注于图 2-7b)。

Solution

Take three points, A, B, and C, as the reference points for calculating σ, u, and σ', as shown in Table 2-1. The values are plotted in Figure 2-7b).

Table 2-1 Calculations for Example 2-1

表 2-1 例题 2-1 的解

Reference Points 参考点	σ(kPa)	u(kPa)	σ'(kPa)
A	3 × 17 = 51.0	0	51.0
B	(3 × 17) + (2 × 20) = 91.0	2 × 9.8 = 19.6	71.4
C	(3 × 17) + (2 × 20) + (4 × 19) = 167.0	6 × 9.8 = 58.8	108.2

例题 2-2

在 300 m 深的海水下方，计算并绘制距离海洋底部 2 m 深处土体的 σ，u 和 σ'，土体重度为 17.5 kN/m³。这种高水压会压缩土体吗？

Example 2-2

Calculate and draw σ, u, and σ' for a soil element at a depth of 2 m from an ocean-bottom surface under 300-m-deep water. The soil's unit weight is 17.5 kN/m³. Does this high water pressure compress soil?

Solution 解：

$$\sigma = H_w\gamma_w + H_s\gamma_s = 300 \times 9.81 + 2 \times 17.5 = 2978(\text{kPa})$$

$$u = H_w\gamma_w = (300 + 2) \times 9.81 = 2963(\text{kPa})$$

$$\sigma' = \sigma - u = 2978 - 2963 = 15(\text{kPa})$$

The total stress and pore water pressure are very high, but the effective stress is very low. Since the formation of the soil's skeleton is controlled by interparticle stress (effective stress), soils at the near surface of the ocean bottom are not compressed much because of the rather small effective overburden stress. Profiles of σ, u, and σ' for a soil element at a depth of 2 m from an ocean-bottom surface under 300-m-deep water are shown in Figure 2-8.

总应力和孔隙水压力很高，但有效应力很低。由于土体骨架的形成受颗粒间应力(有效应力)的控制，海底近表面土体的有效上覆应力很小，因此不会产生明显压缩。水深 300 m 处海底表面以下 2 m 位置的土体总应力 σ、孔隙水压力 u 以及有效应力 σ'分布如图 2-8 所示。

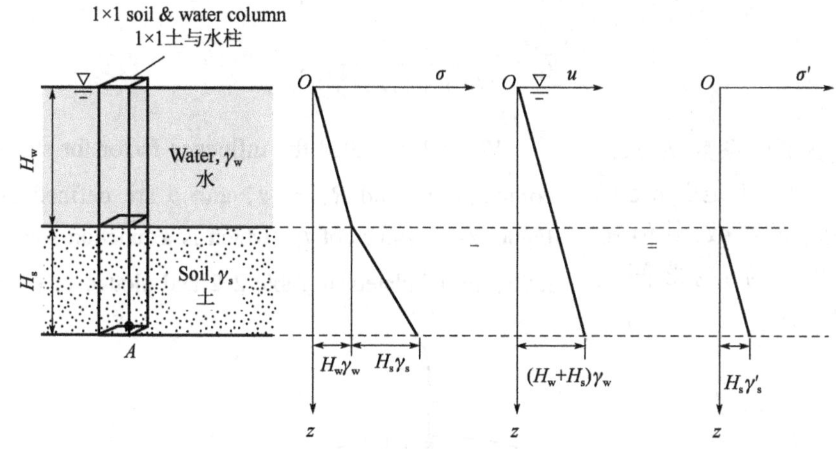

Figure 2-8 Example 2-2 effective stress computation for underwater soil layers

图 2-8 例题 2-2 水下土层有效应力计算

2.4 Additional stresses in a soil mass

When additional loads are placed on a ground surface, such as by footings and traffic loads, those additional loads increase the stresses in the soil mass. These extra stresses are major sources of settlement of soils. The distribution of additional stresses within a soil from applied surface loads or stresses is determined by assuming that the soil is a semi-infinite, homogeneous, linear, isotropic, elastic material. A semi-infinite mass, also called an "elastic half-space", is bounded on one side and extends infinitely in all other directions. For soils, the horizontal surface is the boundary side. Because of the assumption of linear elasticity of the

2.4 土体的附加应力

在地面上施加外部荷载时，例如基础荷载和交通荷载，会引起地基附加应力。附加应力是地基沉降的主要诱因。假定地基土是一种半无限、均匀、线性和各向同性的弹性体，半无限体在一个方向为一平面，其他方向都是无穷大，故也称为"弹性半空间"。对于地基土，水平面为约束边界。由于土体视为线弹性体，可以利用叠加原理对不同荷载作用下土体中某一点在某一方向上的应力增量进行叠加。

soil mass, we can use the principle of superposition. That is, the increases in stress at a given point in a soil mass in a certain direction from different loads can be added together.

基于上述假设,下面给出了几种地表荷载下地基附加应力的计算公式和图表。

Equations and charts are presented for several types of flexible surface loads based on the above assumptions.

2.4.1 点荷载

在半无限弹性体表面某点作用一竖向集中力 P,如图 2-9 所示,弹性体内任一点的应力和位移已由 Boussiuesq 解得(1885)。其中,竖向应力的计算方法为:

2.4.1 Point load

Boussinesq (1885) developed a solution for stresses in an isotropic, homogeneous elastic half-space due to a point load on the surface, as shown in Figure 2-9. The vertical stress increment at a radial distance r from the loading point is given by:

$$\Delta\sigma_v = \frac{3}{2\pi}\frac{P}{z^2}\cos^5\theta = \frac{3Pz^3}{2\pi(r^2+z^2)^{5/2}} = \frac{P}{z^2}\frac{3}{2\pi}\frac{1}{[1+(r/z)^2]^{5/2}} = \frac{P}{z^2}I_1 \tag{2-6}$$

$$I_1 = \frac{3}{2\pi}\frac{1}{[1+(r/z)^2]^{5/2}} \tag{2-7}$$

式中,I_1为影响系数,R、r、z、θ,如图 2-9 所示。I_1是以 r/z 比值为唯一变量的函数,其值在表 2-2 中列出,I_1 与 r/z 之间关系如图 2-10 所示。

Where I_1 is called the influence factor for stress increment computation, and R, r, z, and θ are defined as shown in Figure 2-9. Values of I_1, which is a function solely of the r/z ratio, are tabulated in Table 2-2 and plotted in Figure 2-10.

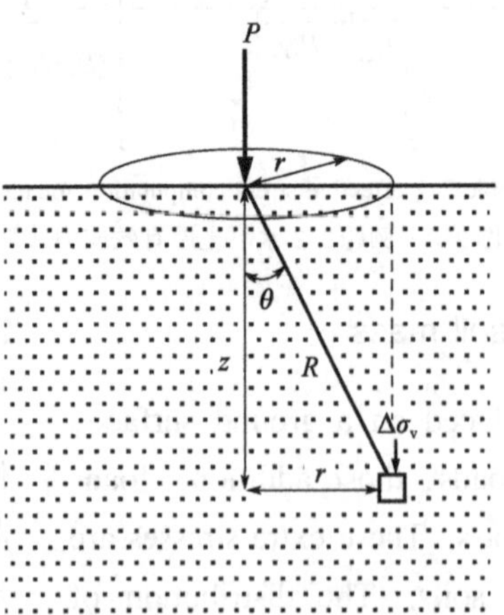

Figure 2-9 Vertical stress increment due to a point load

图 2-9 集中荷载作用时土中竖向附加应力

Table 2-2 Values of influence factor I_1 for a point load

表 2-2 点荷载作用下影响系数 I_1 的取值

r/z	I_1	r/z	I_1	r/z	I_1
0	0.4775	0.04	0.4756	0.08	0.4699
0.02	0.4770	0.06	0.4732	0.10	0.4657

continue
续上表

r/z	I_1	r/z	I_1	r/z	I_1
0.12	0.4607	0.40	0.3295	0.95	0.0956
0.14	0.4548	0.42	0.3181	1.0	0.0844
0.16	0.4482	0.44	0.3068	1.2	0.0513
0.18	0.4409	0.46	0.2955	1.4	0.0317
0.20	0.4329	0.48	0.2843	1.6	0.0200
0.22	0.4243	0.50	0.2733	1.8	0.0129
0.24	0.4151	0.55	0.2466	2.0	0.0085
0.26	0.4054	0.60	0.2214	2.2	0.0058
0.28	0.3954	0.65	0.1978	2.4	0.0040
0.30	0.3849	0.70	0.1762	2.6	0.0028
0.32	0.3742	0.75	0.1565	2.8	0.0021
0.34	0.3632	0.80	0.1386	3.0	0.0015
0.36	0.3521	0.85	0.1226	4.0	0.0004
0.38	0.3408	0.90	0.1083	5.0	0.0001

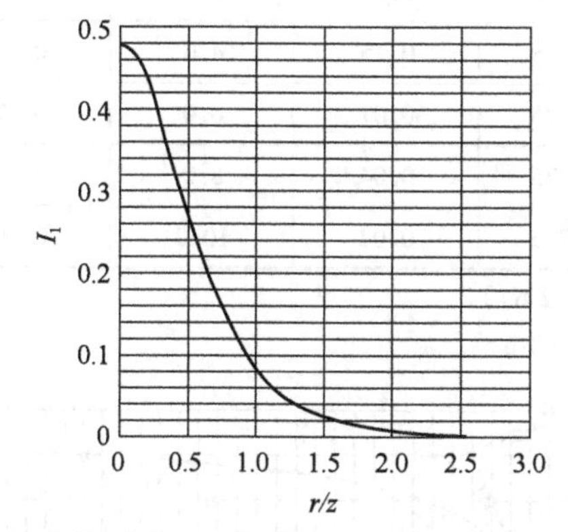

Figure 2-10 Influence factor I_1 versus r/z (point load)

图 2-10 不同 r/z 值对应的影响系数 I_1(点荷载)

Example 2-3

A 5 kN point load is applied on a ground surface. Use Boussinesq's method to compute and plot the magnitudes of the vertical stress increment:

(1) under the point load at depths z from 0 to 10 m below the ground surface;

(2) at a horizontal distance 1.0 m from the load application point at the same depths as above.

例题 2-3

在地面施加一个 5 kN 的点荷载,使用 Boussinesq 解法计算以下工况的竖向附加应力:

(1)点荷载正下方 0 ~ 10 m 深度处的竖向附加应力;

(2)水平距离点荷载 1.0 m,地表以下 0 ~ 10 m 深度处的竖向附加应力。

Solution

(1) For $r/z = 0$, $I_1 = 0.4775$ from Equation (2-7) or Table 2-2.

解:

(1)对于 $r/z = 0$,根据式(2-7)或表 2-2 可知,$I_1 = 0.4775$。

(2)r =1 m, r/z 随深度变化。

通过表 2-3 进行计算,结果如图 2-11 所示。式(2-6)给出了点荷载正下方(即 $r=0$ 和 $z=0$)的 $\Delta\sigma_v$ 值。当距离 r 取非零值时,那么在 $z=0$ 处 $\Delta\sigma_v$ 值为零。

(2)r =1 m, and r/z varies with depth.

Table 2-3 is created for the computation, and the results are plotted in Figure 2-11. Note that Equation (2-6) gives an infinite value of $\Delta\sigma_v$ directly beneath the point load (that is, where $r=0$ and $z=0$) as a special case. When the distance r takes a non-zero value, the value of $\Delta\sigma_v$ becomes zero at $z=0$.

Table 2-3 Values of $\Delta\sigma_v$ under a point load

表 2-3 点荷载作用下 $\Delta\sigma_v$ 计算值

(1) r=0 m				(2) r=1 m			
z(m)	r/z	I_1	$\Delta\sigma_v$	z(m)	r/z	I_1	$\Delta\sigma_v$
0	0	0.4775	∞	0	∞	0	0
0.3	0	0.4775	26.53	0.3	3.33	0.0009	0.05
0.5	0	0.4775	9.55	0.5	2.00	0.0085	0.17
1.0	0	0.4775	2.39	1.0	1.00	0.0844	0.42
2.0	0	0.4775	0.60	2.0	0.50	0.2733	0.34
4.0	0	0.4775	0.15	4.0	0.25	0.4103	0.13
6.0	0	0.4775	0.07	6.0	0.17	0.4459	0.06
8.0	0	0.4775	0.04	8.0	0.13	0.4593	0.04
10.0	0	0.4775	0.02	10.0	0.10	0.4657	0.02

Note: Column $\Delta\sigma_v = P/z^2 \times I_1$ [式(2-6)].

注:土柱 $\Delta\sigma_v = P/z^2 \times I_1$ [Equation (2-6)].

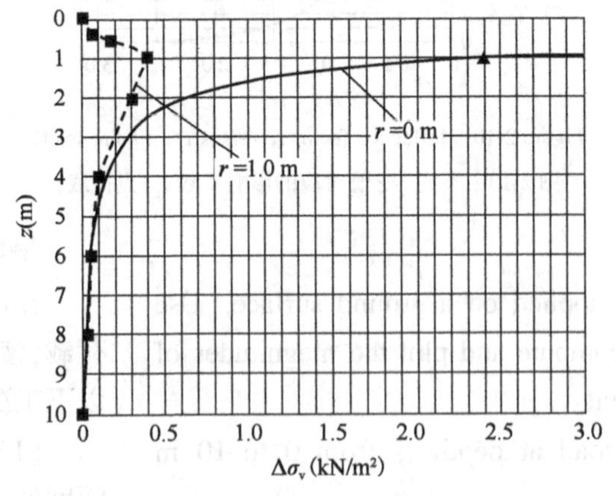

Figure 2-11 $\Delta\sigma_v$ distribution under a point load (Example 2-3)

图 2-11 点荷载作用下 $\Delta\sigma_v$ 分布(例题 2-3)

2.4.2 线荷载

其他形式外荷载的地基附加应力可通过 Boussinesq 点荷载解[式(2-6)]的积分解得。在

2.4.2 Line load

Solutions for stresses under loads applied over lines or areas can be derived by integrating Boussinesq's point load solution [Equation (2-6)] over the line or area over which the load is applied to the ground surface. Figure 2-12 shows a line load q applied to an infinitely long line on the surface of an elastic half-space. Values of $\Delta\sigma_v$ at locations (z, r), where r is the distance measured perpendicular to the line of the load and z is the depth, can be determined by integrating Equation (2-6) over a loaded line from $-\infty$ to $+\infty$, which yields:

地表沿一条无限长直线上连续均匀分布的荷载 q 称为线荷载，如图2-12所示。在这种荷载条件下，地基中任一点的应力状态 $\Delta\sigma_v$ 是 (z,r) 坐标的函数，r 为垂直于线荷载的距离，z 为深度。根据 Boussinesq 解[式(2-6)]，沿荷载长度方向从 $-\infty$ 到 $+\infty$ 积分求得：

$$\Delta\sigma_v = \frac{2qz^3}{\pi R^4} = \frac{2q}{\pi z\left[(r/z)^2+1\right]^2} = \frac{q}{z}\frac{2}{\pi\left[(r/z)^2+1\right]^2} = \frac{q}{z}I_2 \tag{2-8}$$

$$I_2 = \frac{2}{\pi\left[(r/z)^2+1\right]^2} \tag{2-9}$$

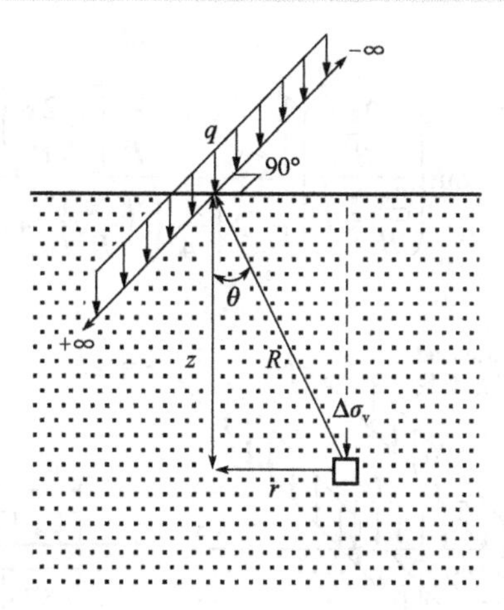

Figure 2-12 Vertical stress increment due to a line load

图2-12 均布线荷载作用时土中竖向附加应力

Values of the influence factor I_2, which is a function solely of the r/z ratio, are tabulated in Table 2-4.

表2-4列出了影响系数 I_2 的值，该值仅为 r/z 的函数。

Table 2-4 Values of influence factor I_2 for line load solution

表2-4 线荷载作用下影响系数 I_2 值

r/z	I_2	r/z	I_2	r/z	I_2
0	0.637	0.8	0.237	1.6	0.050
0.1	0.624	0.9	0.194	1.7	0.042
0.2	0.589	1.0	0.159	1.8	0.035
0.3	0.536	1.1	0.130	1.9	0.030
0.4	0.473	1.2	0.107	2.0	0.025
0.5	0.407	1.3	0.088	2.2	0.019
0.6	0.344	1.4	0.073	2.4	0.014
0.7	0.287	1.5	0.060	2.6	0.011

continue
续上表

r/z	I_2	r/z	I_2	r/z	I_2
2.8	0.008	3.4	0.004	4.0	0.002
3.0	0.006	3.6	0.003	5.0	0.001
3.2	0.005	3.8	0.003		

2.4.3 条形荷载

均布条形荷载指沿长度方向无限延伸且宽度为 B 的均匀分布荷载,如图 2-13 所示。(x,z)点的附加应力 $\Delta\sigma_v$ 通过对式(2-6)沿 x 方向从 $-B/2$ 到 $+B/2$,沿 y 方向从 $-\infty$ 到 $+\infty$ 进行积分求得:

2.4.3 Strip load

A uniformly distributed strip load q with a footing width B applied to an elastic half-space is illustrated in Figure 2-13. Values of $\Delta\sigma_v$ at any point (x, z) can be obtained by integrating Equation (2-6) over x from $-B/2$ to $+B/2$ and over y from $-\infty$ to $+\infty$. The integration solution is given by:

$$\Delta\sigma_v = \frac{q}{\pi}[\beta + \sin\beta\cos(\beta + 2\delta)]$$

$$= \frac{q}{\pi}\left\{\left[\arctan\left(\frac{\frac{2z}{B}}{\frac{2x}{B}-1}\right) - \arctan\left(\frac{\frac{2z}{B}}{\frac{2x}{B}+1}\right)\right] - \frac{\frac{2z}{B}\left[\left(\frac{2x}{B}\right)^2 - \left(\frac{2z}{B}\right)^2 - 1\right]}{2\left[\frac{1}{4}\left\{\left(\frac{2x}{B}\right)^2 + \left(\frac{2z}{B}\right)^2 - 1\right\} + \left(\frac{2z}{B}\right)^2\right]}\right\} = qI_3 \tag{2-10}$$

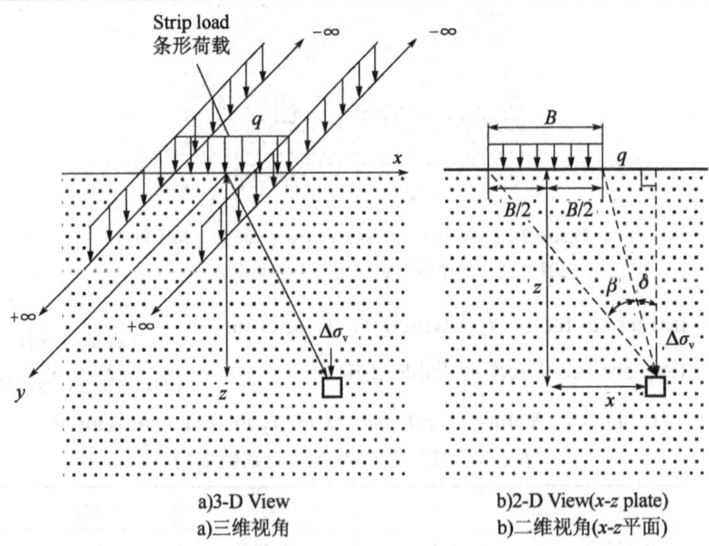

Figure 2-13 Vertical stress increment due to a strip load

图 2-13 均布条形荷载作用时土中竖向附加应力

需要注意的是,在式(2-10)中,当 $2x/B<1$,即点(x,z)在基础以内时,第二行的第一项为负值,此时应加上 π 以解得正值,式(2-10)第二行的第一项替换为:

Note that in Equation (2-10), when $2x/B < 1$ [i. e., when a point (x, z) is within the foundation width B], the value in the first term of the second line becomes negative. To obtain the correct angle value in that case, π should be added to it. That is, the first term of the second line of Equation (2-10) should be replaced as shown below:

$$\arctan\left(\frac{\frac{2z}{B}}{\frac{2x}{B}-1}\right)+\pi \tag{2-11}$$

Table 2-5 shows values of the influence factor I_3 as a function of $2x/B$ and $2z/B$.

表 2-5 给出了以 $2x/B$ 和 $2z/B$ 为变量的 I_3 取值。

Table 2-5 Values of influence factor I_3 for strip load solution

表 2-5 条形荷载作用下影响系数 I_3 值

2z/B	2x/B											
	0	0.2	0.4	0.6	0.8	1	1.25	1.5	2	3	5	10
0	1	1	1	1	1	0.5	0	0	0	0	0	0
0.1	1.000	0.999	0.999	0.997	0.980	0.500	0.011	0.002	0.000	0.000	0.000	0.000
0.2	0.997	0.996	0.992	0.979	0.909	0.500	0.059	0.011	0.002	0.000	0.000	0.000
0.4	0.977	0.973	0.955	0.906	0.773	0.498	0.178	0.059	0.011	0.001	0.000	0.000
0.6	0.937	0.928	0.896	0.825	0.691	0.495	0.258	0.120	0.030	0.004	0.000	0.000
0.8	0.881	0.869	0.829	0.755	0.638	0.489	0.305	0.173	0.056	0.010	0.001	0.000
1.0	0.818	0.805	0.766	0.696	0.598	0.480	0.332	0.214	0.084	0.017	0.002	0.000
1.2	0.755	0.743	0.707	0.646	0.564	0.468	0.347	0.243	0.111	0.026	0.004	0.000
1.4	0.696	0.685	0.653	0.602	0.534	0.455	0.354	0.263	0.135	0.037	0.005	0.000
1.6	0.642	0.633	0.605	0.562	0.506	0.440	0.356	0.276	0.155	0.048	0.008	0.001
1.8	0.593	0.585	0.563	0.526	0.479	0.425	0.353	0.284	0.172	0.060	0.010	0.001
2.0	0.550	0.543	0.524	0.494	0.455	0.409	0.348	0.288	0.185	0.071	0.013	0.001
2.5	0.462	0.458	0.445	0.426	0.400	0.370	0.328	0.285	0.205	0.095	0.022	0.002
3.0	0.396	0.393	0.385	0.372	0.355	0.334	0.305	0.274	0.211	0.114	0.032	0.003
3.5	0.345	0.343	0.338	0.329	0.317	0.302	0.281	0.258	0.210	0.127	0.042	0.004
4.0	0.306	0.304	0.301	0.294	0.285	0.275	0.259	0.242	0.205	0.134	0.051	0.006
5.0	0.248	0.247	0.245	0.242	0.237	0.231	0.222	0.212	0.188	0.139	0.065	0.010
6.0	0.208	0.208	0.207	0.205	0.202	0.198	0.192	0.186	0.171	0.136	0.075	0.015
8.0	0.158	0.157	0.157	0.156	0.155	0.153	0.150	0.147	0.140	0.122	0.083	0.025
10.0	0.126	0.126	0.126	0.126	0.125	0.124	0.123	0.121	0.117	0.107	0.082	0.032
15.0	0.085	0.085	0.085	0.084	0.084	0.084	0.083	0.083	0.082	0.078	0.069	0.041
20.0	0.064	0.064	0.064	0.063	0.063	0.063	0.063	0.063	0.062	0.061	0.056	0.041
50.0	0.025	0.025	0.025	0.025	0.025	0.025	0.025	0.025	0.025	0.025	0.025	0.024
100.0	0.013	0.013	0.013	0.013	0.013	0.013	0.013	0.013	0.013	0.013	0.013	0.012

Example 2-4

A strip load $q = 100$ kPa is applied over a footing with width $B = 5$ m. Calculate and plot the vertical stress distribution under the footing over the x distance at depths of $z = 5$ m and $z = 10$ m.

例题 2-4

将条形荷载 $q = 100$ kPa 施加在宽度 $B = 5$ m 的基础上,计算并绘制基础下 $z = 5$ m 和 $z = 10$ m 处 x 方向上的竖向应力分布。

解:

当$z=5$ m, $2z/B=2\times5/5=2$;

当$z=10$ m, $2z/B=2\times10/5=4$。

对于前述的 $2z/B$ 值,从表2-5中读取对应的I_3值,并在表2-6 中计算不同 x 值对应的$\Delta\sigma_v$值。图 2-14 为半空间($x>0$ 区域)的计算结果。

Solution

At $z=5$ m, $2z/B=2\times5/5=2$;

At $z=10$ m, $2z/B=2\times10/5=4$.

For the preceding $2z/B$ values, I_3 values were obtained from Table 2-5, and $\Delta\sigma_v$ values were computed as shown in Table 2-6 for various x values. The results are plotted in Figure 2-14 for a half-space ($x>0$ region).

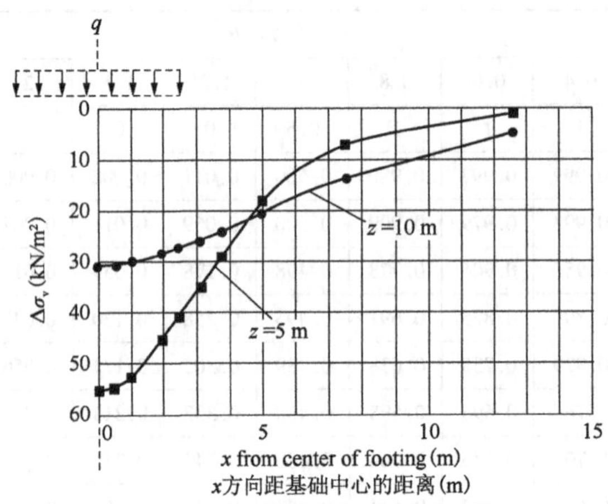

Figure 2-14 Solution for Example 2-4

图2-14 例题2-4 的解

Table 2-6 Computations for Example 2-4

表2-6 例题2-4 的计算

$z=5$ m, $2z/B=2$										
x(m)	0	0.50	1.00	2.00	2.50	3.125	3.75	5.00	7.50	12.50
$2x/B$	0	0.20	0.40	0.80	1.00	1.25	1.50	2.00	3.00	5.00
I_3	0.550	0.543	0.524	0.455	0.409	0.348	0.288	0.185	0.071	0.013
$\Delta\sigma_v$(kN/m²)	55.0	54.3	52.4	45.5	40.9	34.8	28.8	18.5	7.1	1.3
$z=10$ m, $2z/B=4$										
x(m)	0	0.50	1.00	2.00	2.50	3.125	3.75	5.00	7.50	12.50
$2x/B$	0	0.20	0.40	0.80	1.00	1.25	1.50	2.00	3.00	5.00
I_3	0.306	0.304	0.301	0.285	0.275	0.259	0.242	0.205	0.134	0.051
$\Delta\sigma_v$(kN/m²)	30.6	30.4	30.1	28.5	27.5	25.9	24.2	20.5	13.4	5.1

Note: I_3 is from Table 2-5; $\Delta\sigma_v=q\times I_3$.

注:I_3 取自表2-5;$\Delta\sigma_v=q\times I_3$。

2.4.4 圆形均布荷载

圆形基础是一种常见的基础形状,通过 Boussinesq 解对均匀分布的圆形荷载进行积分,如图2-15 所示,可得圆形基础中心正下方 $\Delta\sigma_v$的解,即式(2-12)。

2.4.4 Uniformly loaded circular area

A popular footing shape is a circular one, and Boussinesq's solution can be integrated for a uniformly loaded circular area as shown in Figure 2-15. Equation (2-12) is the solution for $\Delta\sigma_v$ directly under the center of a circular load.

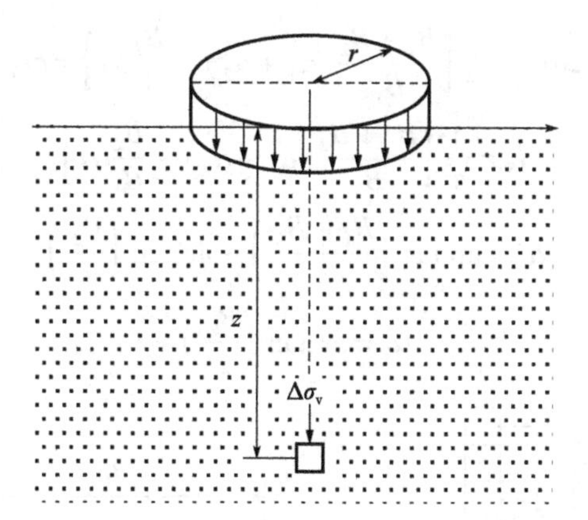

Figure 2-15 $\Delta\sigma_v$ under the center of a circular load

图 2-15 圆形均布荷载作用下土中竖向附加应力

$$\Delta\sigma_v = q\left\{1 - \frac{1}{[(r/z)^2 + 1]^{\frac{3}{2}}}\right\} = qI_4 \tag{2-12}$$

$$I_4 = 1 - \frac{1}{[(r/z)^2 + 1]^{\frac{3}{2}}} \tag{2-13}$$

I_4 values calculated as a function of r/z are tabulated in Table 2-7 and plotted in Figure 2-16.

影响系数 I_4 为以 r/z 为变量的函数，取值见表 2-7，二者关系曲线如图 2-16 所示。

Table 2-7 Values of influence factor I_4 for circular load solution

表 2-7 圆形荷载作用下影响系数 I_4 值

r/z	I_4	r/z	I_4
0	0	1.2	0.738
0.1	0.015	1.4	0.804
0.2	0.057	1.6	0.851
0.3	0.121	1.8	0.885
0.4	0.200	2.0	0.911
0.5	0.284	2.5	0.949
0.6	0.369	3.0	0.968
0.7	0.450	3.5	0.979
0.8	0.524	4.0	0.986
0.9	0.589	4.5	0.990
1.0	0.646	5.0	0.992

2.4.5 Rectangular and triangular loads (embankment loads)

2.4.5 矩形、三角形荷载(路堤荷载)

Some frequently encountered loading patterns are formed due to embankments. Figure 2-17 shows a half section of an embankment load. The integration solution is given by:

路堤荷载也是一种常见的荷载形式。图 2-17 为半幅路堤横截面的荷载分布，计算方法为：

$$\Delta\sigma_{\mathrm{v}} = \frac{q}{\pi}\left[\frac{B_1 + B_2}{B_1}(\alpha_1 + \alpha_2) - \frac{B_2}{B_1}\alpha_2\right] = qI_5 \tag{2-14}$$

$$I_5 = \frac{1}{\pi}\left[\frac{B_1 + B_2}{B_1}(\alpha_1 + \alpha_2) - \frac{B_2}{B_1}\alpha_2\right] \tag{2-15}$$

$$\alpha_1 = \arctan\left(\frac{B_1 + B_2}{z}\right) - \arctan\left(\frac{B_2}{z}\right)$$

$$\alpha_2 = \arctan\left(\frac{B_2}{z}\right) \tag{2-16}$$

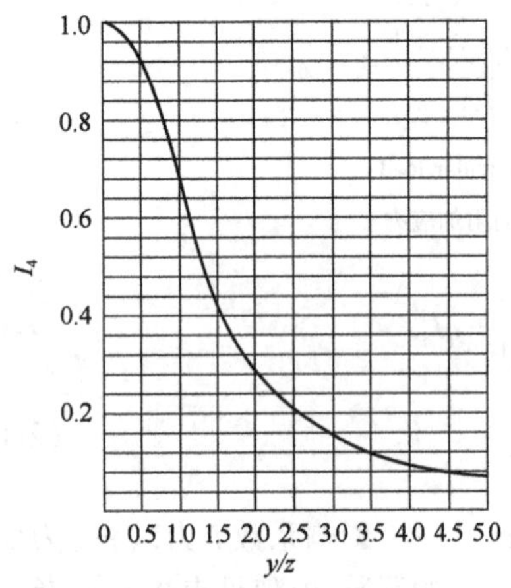

Figure 2-16 Influence factor I_4

图 2-16 影响系数 I_4

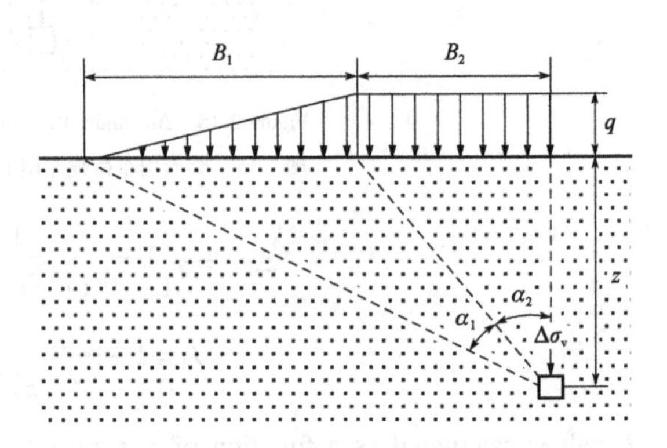

Figure 2-17 Vertical stress increment under a half embankment load

图 2-17 半路堤荷载作用时土中竖向附加应力

影响系数 I_5 为 B_1/z 和 B_2/z 的函数,其值见表 2-8,如图 2-18 所示。路堤下附加应力可通过应力叠加来计算。

Values of the influence factor I_5 as a function of B_1/z and B_2/z are tabulated in Table 2-8 and plotted in Figure 2-18. This convenient solution for the stress increment under embankments is achieved using a superposition of solutions.

Table 2-8 Values of influence factor I_5 for rectangular and triangular load solutions

表 2-8 矩形和三角形荷载作用下影响系数 I_5 值

	B_1/z														
B_2/z	0.01	0.02	0.04	0.06	0.1	0.2	0.4	0.6	0.8	1.0	2.0	4.0	6.0	8.0	10.0
0	0.003	0.006	0.013	0.019	0.032	0.063	0.121	0.172	0.215	0.250	0.352	0.422	0.447	0.460	0.468
0.1	0.066	0.069	0.076	0.082	0.094	0.123	0.176	0.221	0.258	0.288	0.375	0.434	0.455	0.466	0.473
0.2	0.127	0.130	0.136	0.141	0.153	0.179	0.227	0.265	0.297	0.322	0.394	0.444	0.462	0.471	0.477
0.3	0.183	0.186	0.191	0.196	0.206	0.230	0.271	0.303	0.330	0.351	0.411	0.452	0.468	0.476	0.480
0.4	0.233	0.235	0.240	0.245	0.253	0.274	0.308	0.336	0.358	0.375	0.425	0.459	0.472	0.479	0.483
0.5	0.277	0.279	0.283	0.287	0.294	0.311	0.340	0.363	0.381	0.395	0.437	0.466	0.477	0.482	0.486
0.6	0.314	0.316	0.319	0.322	0.329	0.343	0.367	0.386	0.400	0.412	0.446	0.471	0.480	0.485	0.488
0.7	0.345	0.347	0.349	0.352	0.357	0.369	0.389	0.404	0.416	0.426	0.454	0.475	0.483	0.487	0.489
0.8	0.371	0.372	0.375	0.377	0.381	0.391	0.407	0.419	0.429	0.437	0.461	0.478	0.485	0.489	0.491
0.9	0.392	0.393	0.395	0.397	0.401	0.408	0.422	0.432	0.440	0.447	0.467	0.481	0.487	0.490	0.492
1.0	0.410	0.411	0.412	0.414	0.416	0.423	0.434	0.442	0.449	0.455	0.471	0.484	0.489	0.491	0.493
1.2	0.436	0.436	0.437	0.438	0.440	0.445	0.452	0.458	0.463	0.466	0.478	0.488	0.491	0.493	0.495

continue
续上表

B_2/z	B_1/z														
	0.01	0.02	0.04	0.06	0.1	0.2	0.4	0.6	0.8	1.0	2.0	4.0	6.0	8.0	10.0
1.4	0.453	0.454	0.454	0.455	0.456	0.459	0.464	0.469	0.472	0.475	0.483	0.490	0.493	0.495	0.496
1.6	0.465	0.466	0.466	0.467	0.467	0.470	0.473	0.476	0.478	0.480	0.487	0.492	0.495	0.496	0.497
1.8	0.474	0.474	0.474	0.475	0.475	0.477	0.479	0.481	0.483	0.485	0.489	0.494	0.496	0.497	0.497
2.0	0.480	0.480	0.480	0.480	0.481	0.482	0.484	0.485	0.487	0.488	0.491	0.495	0.496	0.497	0.498
3.0	0.493	0.493	0.493	0.493	0.493	0.494	0.494	0.494	0.495	0.495	0.496	0.498	0.498	0.499	0.499
5.0	0.498	0.498	0.498	0.498	0.498	0.498	0.498	0.498	0.499	0.499	0.499	0.499	0.499	0.499	0.499

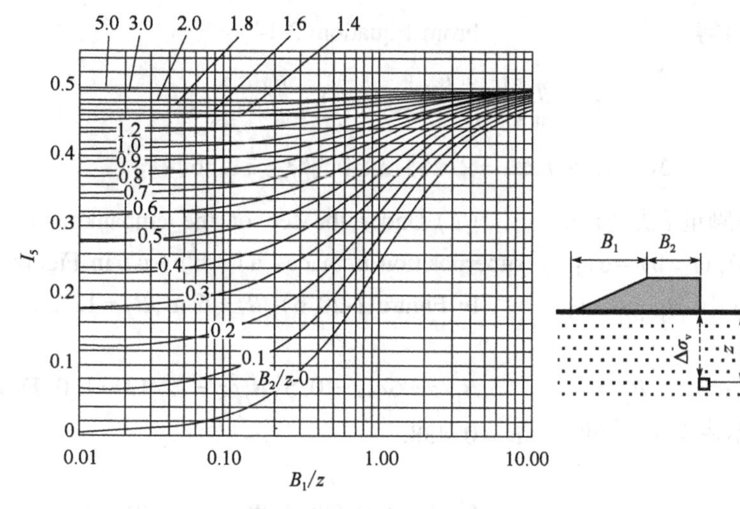

Figure 2-18 Influencing factor I_5

图 2-18 影响系数 I_5

Example 2-5

For the embankment shown in Figure 2-19, determine $\Delta\sigma_v$ at $z = 12$ m below the ground surface:

(1) directly below the centerline of the embankment;

(2) directly under the toe of the embankment.

The total unit weight of the embankment is $\gamma_t = 19.5$ kN/m^3.

例题 2-5

对如图 2-19 所示的路堤，确定地面以下 $z = 12$ m 处 $\Delta\sigma_v$ 值：

(1) 路堤中心线正下方；

(2) 路堤坡脚正下方。

路堤重度为 19.5 kN/m^3。

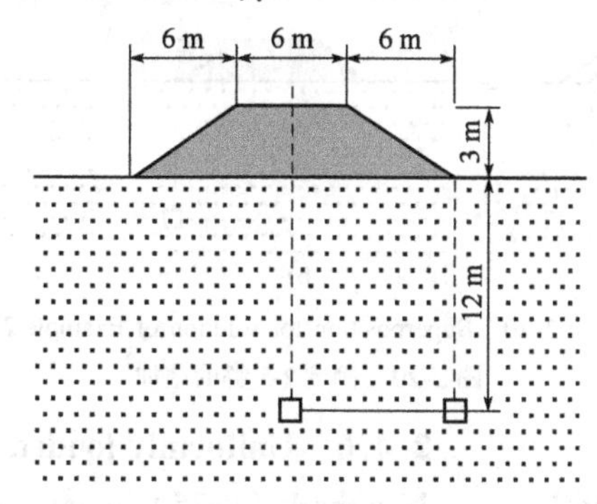

Figure 2-19 Diagram for Example 2-5

图 2-19 例题 2-5 的附图

解：

Solution

$$q=\gamma_t H=19.5\times3=58.5(\text{kPa})$$

(1)在路堤中心线正下方，通过两个相等的半路堤的叠加进行求解。对于半路堤：

(1) For points under the center, the solution is obtained by a superposition of the solutions for two equal half embankments. For each half embankment:

$$B_1=6\ \text{m},B_2=3(\text{m})$$

$$B_1/z=6/12=0.5,B_2/z=3/12=0.25$$

由图2-18可得：

From Figure 2-18:

$$I_5=0.268$$

由式(2-14)可得：

From Equation(2-14):

$$\Delta\sigma_v=\frac{q}{\pi}\left[\frac{B_1+B_2}{B_1}(\alpha_1+\alpha_2)-\frac{B_2}{B_1}\alpha_2\right]=qI_5$$

$$\Delta\sigma_v=2\times q\times I_5=2\times58.5\times0.268=31.36(\text{kPa})$$

(2)在路堤坡脚正下方,叠加求解原理如图2-20,a) = b) - c)：

在图2-20 b)中，$B_1=6$ m，$B_2=12$ m.

$B_1/z=6/12=0.5$，$B_2/z=12/12=1.0$，根据表2-8可知，$I_5=0.438$。

在图2-20 c)中，$B_1=6$ m，$B_2=0$ m。

$B_1/z=6/12=0.5$，$B_2/z=0/12=0$. 根据图2-18可知，$I_5=0.148$。

根据式(2-14)及图2-20b)与c)的叠加可知：

(2) Under the toe of the embankment, the following superposition is made: a) = b) - c) in Figure 2-20.

In Figure 2-20 b), $B_1=6$ m, $B_2=12$ m.

$B_1/z=6/12=0.5$, $B_2/z=12/12=1.0$. From Table 2-8, $I_5=0.438$.

In Figure 2-20 c), $B_1=6$ m, $B_2=0$ m.

$B_1/z=6/12=0.5$, $B_2/z=0/12=0$. From Figure 2-18, $I_5=0.148$.

From Equation (2-14) and by the superposition of two figures [Figure 2-20b) ~ c)]:

$$\Delta\sigma_v=q\times[I_5(\text{b})-I_5(\text{c})]=58.5\times(0.438-0.148)=16.97(\text{kPa})$$

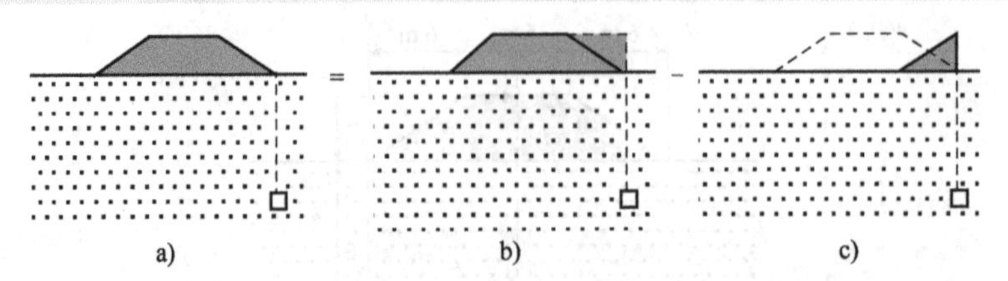

Figure 2-20 Superposition for solution of Example 2-5

图2-20 例题2-5解的叠加

2.4.6 矩形均布荷载

Newmark(1935)对矩形均布荷载作用下的Boussinesq方程

2.4.6 Uniformly loaded rectangular area

Newmark (1935) integrated Boussinesq's equation over a rectangular loading area (Figure 2-21), and the solution under a corner of the footing is given by:

进行了积分，如图 2-21 所示，矩形均布荷载角点下的竖向附加应力为：

$$\Delta\sigma_v = qI_6 \tag{2-17}$$

$$I_6 = \frac{1}{4\pi}\left[\frac{2mn\sqrt{m^2+n^2+1}}{m^2+n^2+m^2n^2+1}\cdot\frac{m^2+n^2+2}{m^2+n^2+1}+\arctan\left(\frac{2mn\sqrt{m^2+n^2+1}}{m^2+n^2-m^2n^2+1}\right)\right] \tag{2-18}$$

Where $m = B/z$ and $n = L/z$. Note that when the arctan(*) term in Equation (2-18) becomes negative, π should be added to that term to obtain the correct I_6 values. Moreover, B and L (or m and n) are interchangeable parameters; thus, B or L can be assigned to either side of a footing.

式中，$m = B/z$，$n = L/z$。需要注意的是，当式(2-18)中的 arctan(*)项变为负值时，应加 π 后重新求解，且 B、L(或 m、n)是可变的，因此基础的任意一侧可被指定为 B 或 L。

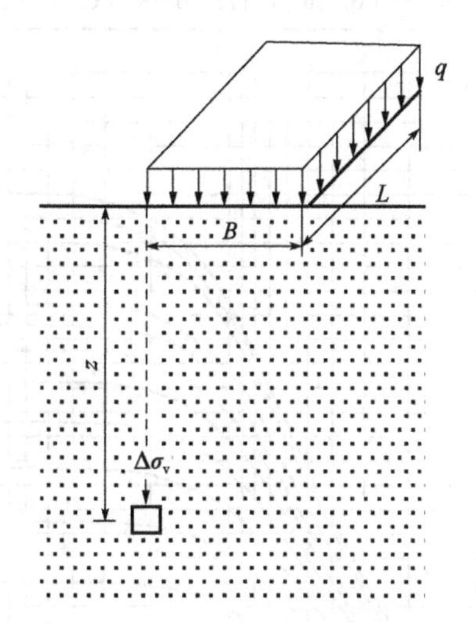

Figure 2-21 $\Delta\sigma_v$ under the corner of rectangular footing

图 2-21 矩形基础角点处的 $\Delta\sigma_v$

Table 2-9 and Figure 2-22 show I_6 values as functions of m and n.

表 2-9 和图 2-22 为以 m 和 n 为变量的影响系数 I_6 的取值。

Table 2-9 Values of influence factor I_6 under corner of rectangular footing

表 2-9 矩形均布荷载基础角点处影响系数 I_6 取值

	m													
n	0.1	0.2	0.3	0.4	0.5	0.6	0.7	0.8	0.9	1.0	1.5	2.0	5.0	10.0
0.1	0.005	0.009	0.013	0.017	0.020	0.022	0.024	0.026	0.027	0.028	0.030	0.031	0.032	0.032
0.2	0.009	0.018	0.026	0.033	0.039	0.043	0.047	0.050	0.053	0.055	0.059	0.061	0.062	0.062
0.3	0.013	0.026	0.037	0.047	0.056	0.063	0.069	0.073	0.077	0.079	0.086	0.089	0.090	0.090
0.4	0.017	0.033	0.047	0.060	0.071	0.080	0.087	0.093	0.098	0.101	0.110	0.113	0.115	0.115
0.5	0.020	0.039	0.056	0.071	0.084	0.095	0.103	0.110	0.116	0.120	0.131	0.135	0.137	0.137

continue
续上表

	m													
n	0.1	0.2	0.3	0.4	0.5	0.6	0.7	0.8	0.9	1.0	1.5	2.0	5.0	10.0
0.6	0.022	0.043	0.063	0.080	0.095	0.107	0.117	0.125	0.131	0.136	0.149	0.153	0.156	0.156
0.7	0.024	0.047	0.069	0.087	0.103	0.117	0.128	0.137	0.144	0.149	0.164	0.169	0.172	0.172
0.8	0.026	0.050	0.073	0.093	0.110	0.125	0.137	0.146	0.154	0.160	0.176	0.181	0.185	0.185
0.9	0.027	0.053	0.077	0.098	0.116	0.131	0.144	0.154	0.162	0.168	0.186	0.192	0.196	0.196
1.0	0.028	0.055	0.079	0.101	0.120	0.136	0.149	0.160	0.168	0.175	0.194	0.200	0.204	0.205
1.5	0.030	0.059	0.086	0.110	0.131	0.149	0.164	0.176	0.186	0.194	0.216	0.224	0.230	0.230
2.0	0.031	0.061	0.089	0.113	0.135	0.153	0.169	0.181	0.192	0.200	0.224	0.232	0.240	0.240
5.0	0.032	0.062	0.090	0.115	0.137	0.156	0.172	0.185	0.196	0.204	0.230	0.240	0.249	0.249
10.0	0.032	0.062	0.090	0.115	0.137	0.156	0.172	0.185	0.196	0.205	0230	0.240	0.249	0.250

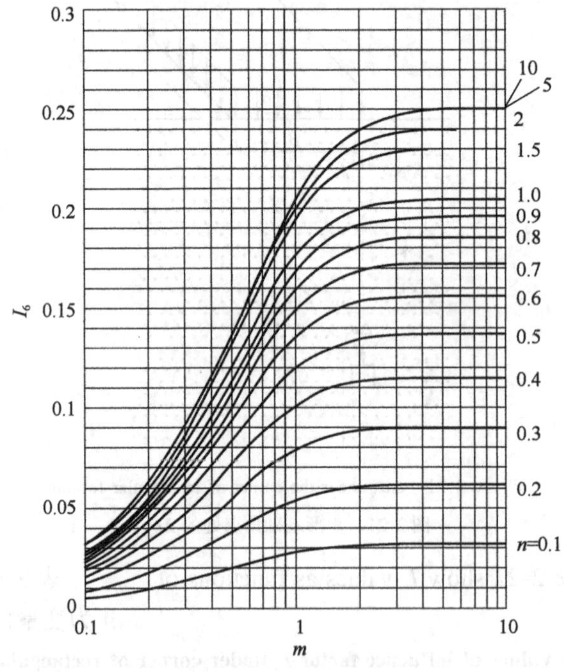

Figure 2-22 Influence factor I_6

图 2-22 影响系数 I_6

式(2-17)为矩形均布荷载角点下的附加应力解。利用叠加原理,该解可用于计算矩形基础任意点下的$\Delta\sigma_v$,如图2-23所示,其中深色区为实际受荷区域,虚线区为假想受荷区域。

The solution in Equation (2-17) is for $\Delta\sigma_v$ under a corner of a rectangular footing. However, the solution can be used to compute $\Delta\sigma_v$ under any point of a rectangular footing, using the principle of superposition. Figure 2-23 shows $\Delta\sigma_v$ values under various points of footings. Real loaded footing areas are shown with a darker color, and imaginary footing sections are drawn with dotted lines.

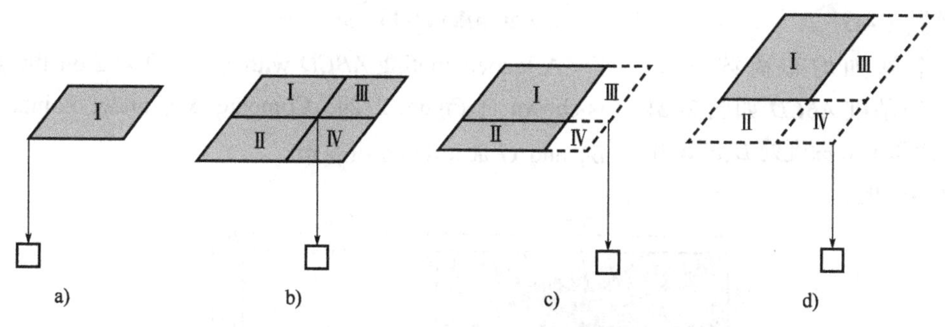

Figure 2-23 Computations for $\Delta\sigma_v$ under various points of footings

图 2-23 计算 $\Delta\sigma_v$在不同点的地基附加应力

Case a) in Figure 2-23 represents an area under a corner of a footing, Case b) under the midpoint of the footing, and Cases c) and d) under points outside the footing. To bring the point of computation to a corner of the footing, including imaginary sections, the following procedures are used:

图 2-23 中示例 a)为矩形均布荷载的角点下方,示例 b)为矩形均布荷载的中点下方,示例 c)和示例 d)在荷载的外侧。计算原理为将计算点移至受荷区的一角,利用叠加原理计算,这一计算方法称为角点法,步骤如下:

Case a): loaded area = Ⅰ; Equation (2-17) is used directly.

示例 a):加载区 = Ⅰ,可直接用式(2-17)计算。

Case b): Loaded areas = Ⅰ + Ⅱ + Ⅲ + Ⅳ.

示例 b):加载区 = Ⅰ + Ⅱ + Ⅲ + Ⅳ。

$$\Delta\sigma_v(\text{Ⅰ}+\text{Ⅱ}+\text{Ⅲ}+\text{Ⅳ}) = \Delta\sigma_v(\text{Ⅰ}) + \Delta\sigma_v(\text{Ⅱ}) + \Delta\sigma_v(\text{Ⅲ}) + \Delta\sigma_v(\text{Ⅳ}) \tag{2-19}$$

Case c): Loaded areas = Ⅰ + Ⅱ.

示例 c):加载区 = Ⅰ + Ⅱ。

$$\Delta\sigma_v(\text{Ⅰ}+\text{Ⅱ}) = \Delta\sigma_v(\text{Ⅰ}+\text{Ⅲ}) + \Delta\sigma_v(\text{Ⅱ}+\text{Ⅳ}) - \Delta\sigma_v(\text{Ⅲ}) - \Delta\sigma_v(\text{Ⅳ}) \tag{2-20}$$

Case d): Loaded area = Ⅰ.

示例 d):加载区 = Ⅰ。

$$\Delta\sigma_v(\text{Ⅰ}) = \Delta\sigma_v(\text{Ⅰ}+\text{Ⅱ}+\text{Ⅲ}+\text{Ⅳ}) - \Delta\sigma_v(\text{Ⅱ}+\text{Ⅳ}) - \Delta\sigma_v(\text{Ⅲ}+\text{Ⅳ}) + \Delta\sigma_v(\text{Ⅳ}) \tag{2-21}$$

In the preceding expression, for example, $\Delta\sigma_v$(Ⅰ + Ⅱ) means the stress increment computation due to the combined footing areas Ⅰ and Ⅱ. In this approach to the problem, all computation points are located at the corners of combined or single footings, and Equation (2-17) is applicable. In case d), the footing Ⅳ is included in the footings (Ⅱ + Ⅳ) and (Ⅲ + Ⅳ) and subtracted twice. Thus, $\Delta\sigma_v$(Ⅳ) is added once. Note that for each real or imaginary footing, the B and L values are different, and different I_6 values should be obtained for all of those footings.

在前述式中,$\Delta\sigma_v$(Ⅰ+Ⅱ)表示矩形均布荷载Ⅰ和Ⅱ组合引起的附加应力,这样计算点位于组合或单个受荷区域的角点处,式(2-17)即可适用。在示例 d)中,受荷区Ⅳ包含在受荷区(Ⅱ+Ⅳ)和(Ⅲ + Ⅳ)中,已减去两次,故应再加上 $\Delta\sigma_v$(Ⅳ)。注意,对于各实际受荷区或假想受荷区,B 和 L 值是不同的,相应地应取不同的 I_6值。

例题 2-6

一个施加均布荷载 q = 200 kPa的基础 $ABCD$ 如图 2-24 所示。计算 5 m 深度处 E、F、B 和 G 点的 $\Delta\sigma_v$ 值。

Example 2-6

A loaded footing $ABCD$ with q = 200 kPa on the ground is shown in Figure 2-24. Compute $\Delta\sigma_v$ under Points E, F, B, and G at a depth of 5 m.

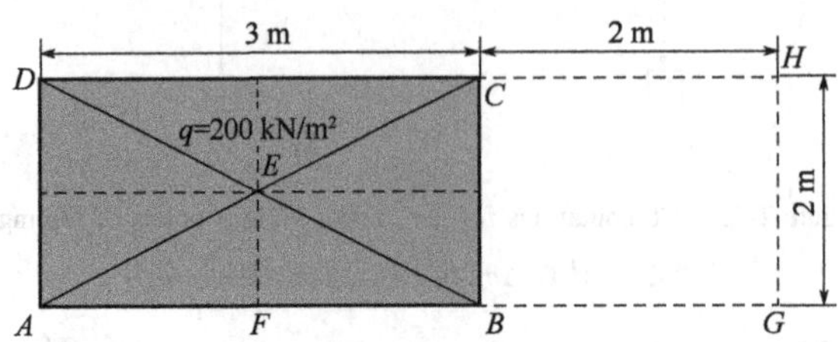

Figure 2-24 Diagram for Example 2-6

图 2-24 例题 2-6 的附图

解:

(1)在 E 点,E 点为四个相等的荷载区域的角点。

B = 1.5 m 且 L = 1 m,于是有:

Solution

(1) At point E, there are four equally sized footings with point E at a corner.

B = 1.5 m and L = 1 m; thus:

$$m = B/z = 1.5/5 = 0.3, n = L/z = 1/5 = 0.2$$

从图 2-22 可知,I_6 = 0.026。根据式(2-17)可知:

From Figure 2-22, I_6 = 0.026. From Equation (2-17):

$$\Delta\sigma_v = 4 \times qI_6 = 4 \times 200 \times 0.026 = 20.8(\text{kPa})$$

(2)在 F 点,F 点为两个相等的荷载区域的角点。

B = 1.5 m 且 L = 2 m,于是有:

(2) At point F, there are two equal footings with point F at a corner.

B = 1.5 m and L = 2 m, thus:

$$m = B/z = 1.5/5 = 0.3, n = L/z = 2/5 = 0.4$$

从图 2-22 可知,I_6 = 0.047。根据式(2-17)可知:

From Figure 2-22, I_6 = 0.047. From Equation (2-17):

$$\Delta\sigma_v = 2 \times qI_6 = 2 \times 200 \times 0.047 = 18.8(\text{kPa})$$

(3)B 点仅是一个荷载区域的角点。

B = 3 m 且 L = 2 m,于是有:

(3) Point B is directly under a corner of footing.

B = 3 m and L = 2 m, thus:

$$m = B/z = 3/5 = 0.6, n = L/z = 2/5 = 0.4$$

从图 2-22 可知,I_6 = 0.080。根据式(2-17)可知:

From Figure 2-22, I_6 = 0.080. From Equation (2-17):

$$\Delta\sigma_v = qI_6 = 200 \times 0.008 = 16.0(\text{kPa})$$

(4)G 为两个假想的荷载区域($AGHD$ 和 $BGHC$)的角点。

(4) At point G, there are two imaginary footings ($AGHD$ and $BGHC$), both of which have point G corners.

For *AGHD*, $B=5$ m and $L=2$ m, thus:

对于 *AGHD* 有,$B=5$ m 且 $L=2$ m,于是有:

$$m=B/z=5/5=1.0, n=L/z=2/5=0.4$$

From Figure 2-22, $I_6=0.101$. For *BGHC*, $B=2$ m and $L=2$ m, thus:

从图 2-22 可知,$I_6=0.101$。对于 *BGHC* 有,$B=2$ m 且 $L=2$ m,于是有:

$$m=B/z=2/5=0.4, n=L/z=2/5=0.4$$

From Figure 2-22, $I_6=0.060$. From Equation (2-17):

从图 2-22 可知,$I_6=0.060$。根据式(2-17)可知:

$$\Delta\sigma_v(ABCD)=AG,(AGHD)-AC,(BGHC)=q\sum I_6=200\times(0.101-0.060)=8.2(\text{kPa})$$

Example 2-7

例题 2-7

There are two adjacent foundations *A* and *B*. Their sizes, positions, and the additional stress distributions are shown in Figure 2-25. Considering the effect of the adjacent foundation *B*, determine the additional stress at a depth *z* of 2 m under the center point *O* of foundation *A*.

有两个相邻的基础 *A* 和 *B*,它们的尺寸、位置和附加应力分布如图 2-25 所示。考虑相邻地基 *B* 的影响,求解基础 *A* 的中心点 *O* 下 2 m 深度 *z* 处的附加应力。

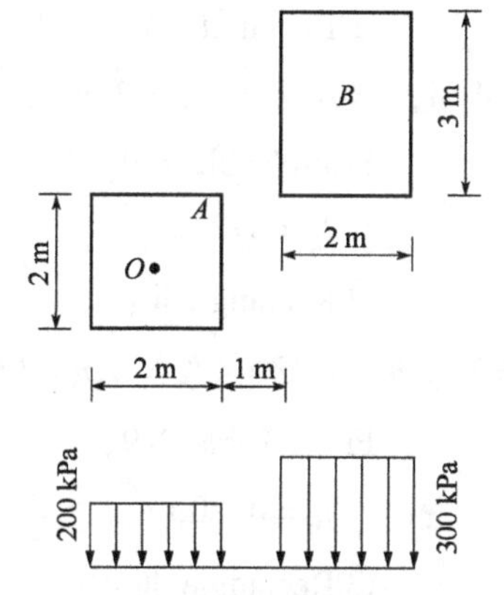

Figure 2-25 Diagram for Example 2-7

图 2-25 例题 2-7 的附图

Solution

(1) Step 1 Determine the additional stress at the center point *O* due to the vertical uniform load of the foundation *A*.

To determine the additional stress under point *O*, the underside of foundation *A* is divided into four rectangles of equal area of 1 m × 1 m. The sum of the additional stresses for the four rectangles is the additional stress at the center point *O* of the foundation *A*, namely, $\sigma_z=4\alpha_c P_A$.

With $m=B/z=1/2=0.5$ and $n=L/z=1/2=0.5$. From Table 2-9, we obtain $I_6=0.0840$. Therefore, the additional

解:

(1)步骤 1:基础 *A* 竖向均布荷载作用下中心点 *O* 的附加应力。

为了求出 *O* 点处的附加应力,地基下侧分为四个等截面 1 m × 1 m的矩形,四个矩形的附加应力之和为地基 *A* 中心点 *O* 处的附加应力,即:

当 $m=B/z=1/2=0.5$ 且 $n=L/z=1/2=0.5$ 时,从表 2-9

可知,$I_6=0.084$. 因此 O 点下的附加应力为:

stress under point O is:

$$\Delta\sigma_v=4\alpha_c P_A=4\times0.084\times200=67.2(\text{kPa})$$

(2)步骤2:中心点 O 的附加应力是由于地基 B 的竖向均匀荷载引起的,根据 O 点在地基 B 外侧来计算附加应力,如图2-26所示,即:

(2) Step 2 Determine the additional stress at the center point O due to the vertical uniform load of the foundation B. Because point O is outside the foundation B, as shown in Figure 2-26, the additional stress is calculated as follows:

$$\Delta\sigma_v=(I_6\text{Ⅰ}-I_6\text{Ⅱ}-I_6\text{Ⅲ}+I_6\text{Ⅳ})P_B$$

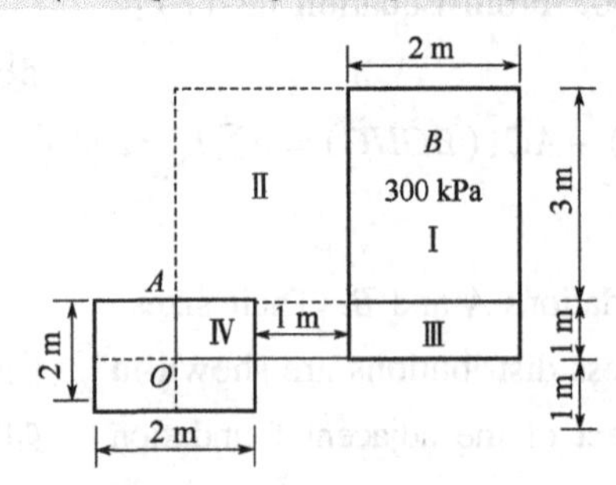

Figure 2-26 Location of center point O of foundation A with respect to foundation B

图2-26 基础 A 中心点 O 相对于地基 B 的位置

①矩形 Ⅰ:

①Rectangle Ⅰ:

$$B=4\text{ m},L=4\text{ m},m=B/Z=4/2=2,n=L/Z=4/2=2$$

从表2-9可知:

From Table 2-9,

$$I_6=0.232$$

②矩形Ⅱ:

②Rectangle Ⅱ:

$$B=4\text{ m},L=2\text{ m},m=B/Z=4/2=2,n=L/Z=2/2=1$$

从表2-9可知:

From Table 2-9,

$$I_6=0.200$$

③矩形Ⅲ:

③Rectangle Ⅲ:

$$B=4\text{ m},L=1\text{ m},m=B/Z=4/2=2,n=L/Z=1/2=0.5$$

从表2-9可知:

From Table 2-9,

$$I_6=0.135$$

④矩形Ⅳ:

④Rectangle Ⅳ:

$$B=2\text{ m},L=1\text{ m},m=B/Z=2/2=1,n=L/Z=1/2=0.5$$

$$\Delta\sigma_v=(I_6\text{Ⅰ}-I_6\text{Ⅱ}-I_6\text{Ⅲ}+I_6\text{Ⅳ})P_B=(0.232-0.200-0.135+0.120)\times300=5.1(\text{kPa})$$

从表2-9可知:

From Table 2-9,

$$I_6=0.120$$

(3)步骤3:在基础 A 以下2 m处的中心点 O 处的附加应力计算如下:

(3) Step 3 The additional stress at the center point O at a depth of 2 m below foundation A is calculated as follows:

$$\Delta\sigma_v = 67.2 + 5.1 = 72.3(\text{kPa})$$

2.5 Contact pressure between foundation and soil

Contact pressure is the intensity of loading transmitted from the underside of a foundation to the ground soil. For the ground soil, the contact pressure is an external load and is the basis for calculating the additional stress in the soil. For the foundation, the contact pressure is the reaction force between the foundation and the ground soil and is the load to be considered in the foundation design. Therefore, it is necessary to study the distribution law and calculation method of the contact pressure, whether calculating the additional stress in the ground soil or designing the foundation structure.

2.5.1 Contact pressure distribution

The magnitude and distribution pattern of the contact pressure depend on many factors, such as the magnitude and distribution of the structure load applied, the rigidity and embedment depth of the foundation, and the soil properties, among others.

Tests have shown that for a foundation with a very low rigidity or for a flexible foundation, the magnitude and distribution pattern of the contact pressure are the same as those of the load applied to the foundation. This is because the deformation of the foundation contact area is comparable to the deformation of the ground soil. When the load on the foundation is uniformly distributed, the contact pressure is also uniformly distributed, as shown in Figure 2-27a). When the load distribution is trapezoidal, the contact pressure distribution is also trapezoidal, as shown in Figure 2-27b).

2.5 基底压力

建筑物荷载通过基础传递给地基,在基础底部与地基之间产生的法向压力称为基底压力。对地基来说,基底压力属于外部荷载,是计算地基中附加应力的依据;对基础来说,基底压力是地基的反作用力,是基础结构设计的荷载。因此,不论计算地基中的附加应力还是基础结构设计,都必须研究基底压力的分布规律和计算方法。

2.5.1 基底压力分布

基底压力的大小和分布形式受到多种因素的影响,包括基础上部荷载的大小和分布、基础的刚度和埋深,以及土体性质等。

大量的实践表明,对于刚度较低的基础或柔性基础,地基与基础的变形是协同发生的,因此基底压力的大小和分布模式与施加在基础上部的荷载相同。当在基础上施加均布荷载时,基底压力也均匀分布,如图 2-27a)所示;当施加梯形荷载时,基底压力也呈梯形分布,如图 2-27b)所示。

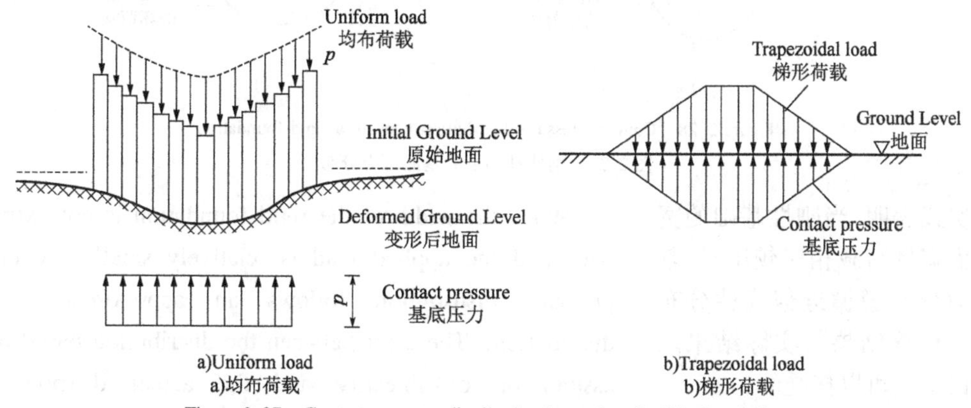

Figure 2-27 Contact pressure distribution beneath a flexible foundation

图 2-27 柔性基础下的基底压力分布

对于地基和基础刚度差异较大的刚性基础,基底压力的分布随所施加荷载的大小、基础埋置深度和地基土性质而变化。例如,在砂土地基中的刚性条形基础上施加中心荷载时,基础中心处的基底压力最大,基础边缘处的基底压力为零,其分布模式类似于抛物线,如图 2-28a)所示,原因在于砂粒没有黏聚作用。但是当在黏土地基中的刚性条形基础上施加中心荷载时,由于黏性土的黏结力,基础边缘也可以承担一部分荷载。当施加的荷载相对较小时,基础边缘处压力较大,而基础中心处压力较小,形成类似马鞍形的压力分布曲线。当荷载逐渐增加到接近破坏荷载时,基础中心处压力随之增大,而基础边缘处压力减小,类似碗状,如图 2-28b)所示。

For a rigid foundation that cannot deform in a manner compatible with the ground deformation because of a significant difference in rigidity, the distribution of the contact pressure varies with the magnitude of the applied load, the embedment depth of the foundation, and the properties of the ground soil. For instance, when a point load is applied to a rigid strip foundation on the surface of a sand, the contact pressure is the greatest at the centerline of the foundation and zero at the edge of the foundation. The contact pressure distribution takes the shape of a parabolic curve, as shown in Figure 2-28a), because of the lack of cohesion among sand particles. When a point load is applied to a rigid strip foundation resting on the surface of a clay, some degree of load can be supported at the edge of the foundation because of the cohesion among clay particles. Therefore, when the applied load is relatively small, high contact pressure is imposed at the edge of the foundation, and low contact pressure is applied at the center of the foundation. The distribution curve has a saddle shape. When the load is increased gradually to the failure load level, the distribution curve of the contact pressure rises at the center and falls at the edge of the foundation, similar to a bowl shape, as shown in Figure 2-28b).

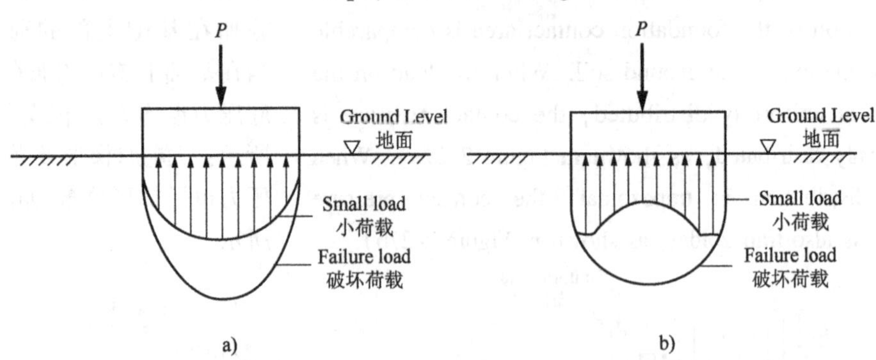

Figure 2-28 Contact pressure distribution beneath a rigid foundation

图 2-28 刚性基础下的基底压力分布

实践表明,当刚性基础的宽度不太宽且荷载相对较小时,基底压力分布遵循近似线性分布假设。计算结果与实际结果之间的误差是可以接受的。

When the width of the rigid foundation is not extremely wide and the applied load is relatively small, the contact pressure distribution follows an approximately linear distribution. The error between the distribution based on the assumption of linearity and the actual distribution is acceptably small.

2.5.2 Simplified calculation of contact pressure

The simplified calculation method normally used in engineering practice for computing contact pressure based on the linear distribution assumption is introduced below.

1) Contact pressure due to vertical centric load

The length and width of a rectangular foundation are l and b, respectively, as shown in Figure 2-29a) and 2-29b). A vertical centric load F is applied to the foundation. According to the linear distribution assumption, the value of the contact pressure is:

2.5.2 基底压力的简化计算

下面介绍工程实践中常用的基于线性分布假设的基底压力简化计算方法。

1) 中心荷载引起的基底压力

矩形基础的长度和宽度分别为 l 和 b,如图 2-29a) 和 2-29b) 所示。在基础上施加垂直中心荷载 F。根据线性分布假设,基底压力值为:

$$p=\frac{F+G}{A}=\frac{F+G}{l\times b} \tag{2-22}$$

Where p represents the contact pressure (kPa); F represents the vertical load on the upper side of the foundation (kN); G represents the self-weight of the foundation and the soil weight on the steps of foundation (for which an average unit weight of 20 kN/m^3 is typically adopted); and $A=l\times b$ represents the area of the foundation (m^2), where l and b represent the length and width of the foundation, respectively.

式中,p 为基底压力(kPa);F 为基础上部竖向荷载(kN);G 为基础自重和基础上部的土体重量,一般取值为 20 kN/m^3;$A=l\times b$ 为基础底面积(m^2),l 和 b 为基础的长度和宽度。

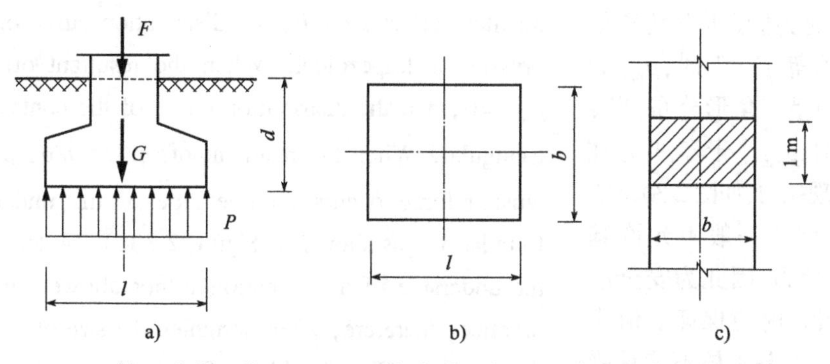

Figure 2-29 Contact pressure distribution due to vertical centric load

图 2-29 中心荷载作用下的基底压力分布

If the foundation is oblong (theoretically, when l/b approaches infinity, it is called a strip foundation; practically, when l/b is greater than or equal to 10, it can be considered a strip foundation), a free body with a 1-m length can be truncated in the longitudinal direction of the foundation for the purpose of calculation analysis, as shown in Figure 2-29c). In this situation, the contact pressure is:

如果基础是条形的(理论上当 l/b 接近无穷大时,称为条形基础。实际上当 l/b 大于或等于 10 时,即可视为条形基础),可以在基础纵向上截取一个 1 m 单位长度的隔离体进行计算分析,如图 2-29c) 所示。在这种情况下,基底压力为:

$$p = \frac{F+G}{b} \quad (2\text{-}23)$$

式中,b 为基础宽度(m);其他符号与前述相同。

Where b represents the width of the foundation (m) and the other symbols have the same meanings as before.

2)偏心荷载引起的基底压力

2)Contact pressure due to one-way vertical eccentric load

当在矩形基础上施加单向偏心荷载时,如图2-30所示,可以用材料力学中的偏心受压公式计算任意点处的基底压力:

When a one-way eccentric load is applied to a rectangular foundation, as shown in Figure 2-30, the contact pressure at any point can be calculated using the following formula for eccentric compression in materials:

$$p_{\min}^{\max} = \frac{F+G}{A} \pm \frac{M}{W} \quad (2\text{-}24)$$

式中,p_{max}和p_{min}为基础的最大和最小基底压力(kPa);M代表Y-Y轴偏心荷载的力矩;W为基础底部的抵抗力矩,如果基础截面积为矩形,则$W=bl^2/6(m^3)$;e是偏心荷载线相对于Y-Y轴的偏移。

Where p_{max} and p_{min} are the maximum and minimum contact pressures, respectively, on both sides of the underside of the foundation (kPa); M is the moment of the eccentric load about the Y-Y axis; W is the resisting moment of the underside of foundation, $W=bl^2/6(m^3)$ if the area is rectangular; and e is the offset of the eccentric load line from the Y-Y axis.

由式(2-24)可以看出,当e值小于l/b时,基底压力呈梯形分布;当e值等于l/b时,p_{min}为零,基底压力呈三角形分布;当e值大于l/b时,p_{min}小于零,并且在基底一侧出现拉力,如图2-30c)所示。工程实践中一般不允许基础底部出现拉力,因此为安全起见,在基础设计时应保证e值小于1/6。由于地基土体不能承受拉力,基底压力的数值也会相应调整。计算原理为基底压力的合力与总荷载相同[图2-30d)],最大基底压力p_{max}计算公式为:

From Equation (2-24), it is understood that when the resultant offset $e<l/b$, the distribution curve of the contact pressure is trapezoidal. When the resultant offset $e=l/6$, $p_{min}=0$, and the distribution curve of the contact pressure is triangular. When the resultant offset $e>l/6$, $p_{min}<0$, and tension force occurs on one side of the underside of the foundation, as shown in Figure 2-30c). A tension force on the underside of a foundation is not allowed in engineering practice; therefore, when designing the size of the foundation, the resultant offset should satisfy the criterion that e is less than 1/6, for the sake of safety. Because soil cannot bear tension force, the contact pressure is adjusted. The calculation principle is that the composite foundation base pressure is identical to the total load [as shown in Figure 2-30d)]. The formula for the maximum contact pressure p_{max} is given by:

$$p_{max} = \frac{2(F+G)}{3ba} \quad (2\text{-}25)$$

Where a is the distance between the application point of the eccentric load and the edge at which the contact pressure is p_{max}, i. e., $a = (l/2 - e)$ m.

式中,a 为偏心荷载作用点与 p_{max} 的距离,$a = (l/2 - e)$ m。

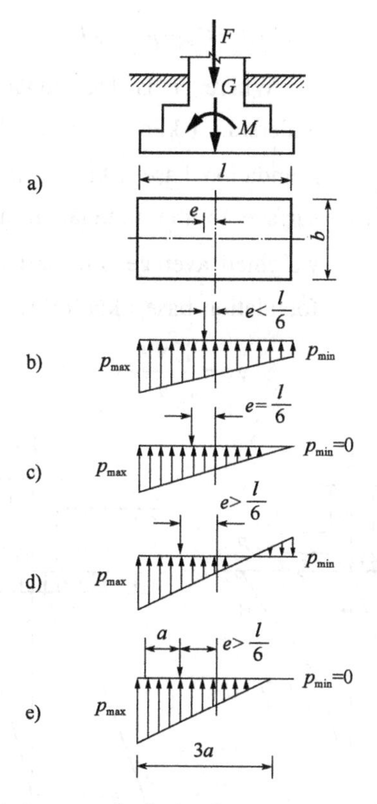

Figure 2-30 Contact pressure distribution due to one-way vertical eccentric load

图 2-30 单向偏心荷载作用下的基底压力分布

Similarly for a strip foundation, the maximum and minimum contact pressures on the underside of the foundation are given by:

类似地,对于条形基础,基础底部的最大和最小基底压力为:

$$p_{\min}^{\max} = \frac{F+G}{A}\left(1 \pm \frac{6e}{b}\right) \tag{2-26}$$

2.5.3 Additional stress on the foundation underside

The additional stress is defined as the increased pressure in the foundation due to the building architecture, as shown in Figure 2-31.

(1) When the foundation is constructed above the ground surface [Figure 2-31a)], the additional stress on the foundation underside p_0 is the contact pressure on the foundation p, that is:

2.5.3 基底附加压力计算

基底附加压力指建筑物建成后使基础底面净增加的压力,如图 2-31 所示。

(1) 当基础位于地面之上[图 2-31a)]时,基底附加压力 p_0 为基底压力 p,即:

$$p_0 = p \tag{2-27}$$

(2)当基础位于地面以下某个深度处[图2-31b)]时,基底附加压力p_0的计算方法为:

(2) When the foundation is constructed at some depth below the ground surface [Figure 2-31b)], the additional stress on the foundation underside p_0 is calculated as follows:

$$p_0 = p - \sigma_c = p - \gamma_0 d \tag{2-28}$$

式中,p为基底压力(kPa);σ_c为基底的上覆层压力(kPa);d为从地面到基底的深度(m);γ_0为基底以上土层的加权平均单位重度(kN/m^3)。

Where p is the contact pressure on the foundation underside (kPa); σ_c is the overburden pressure at the foundation base (kPa); d is the depth from the ground surface to the foundation underside (m), and γ_0 is the weighted average unit weight of the soil layers above the foundation base (kN/m^3).

$$\gamma_0 = \frac{\sum \gamma_i h_i}{d} \tag{2-29}$$

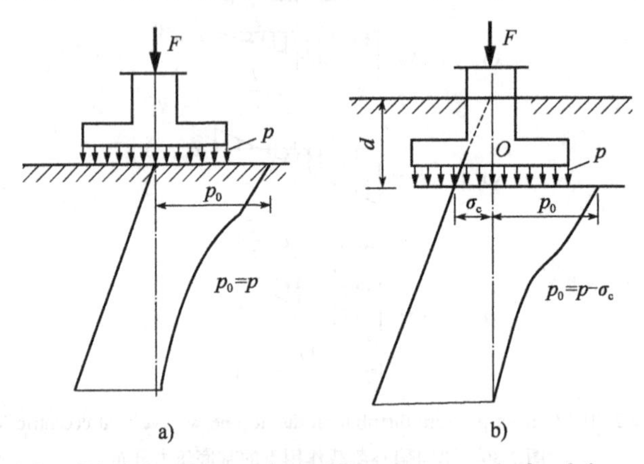

Figure 2-31 Calculation diagram of the additional stress on the foundation underside

图2-31 基底附加压力的计算

在计算地基附加压力时应以p_0为依据,即把p_0视为作用在弹性半无限体表面的局部荷载来考虑。

The additional stress on the ground soil should be calculated based on p_0, that is, p_0 should be considered as the local load acting on the surface of an elastic semi-infinite body.

习 题

(1)土层如图2-32所示:①试绘出自重应力沿深度分布曲线;②当地下水位处于20 m高程时,绘出自重应力沿深度分布曲线。

(2)均布荷载$p=250$ kPa,作用于如图2-33所示的阴影部分,求A点以下3 m深度处的竖向应力。

Exercises

(1) For the soil layer shown in Figure 2-32. ①draw the distribution curve of the self-weight stress along the depth, and ②draw the distribution curve of self-weight stress along the depth when the groundwater level is at an elevation of 20 m.

(2) A uniformly distributed pressure $p = 250$ kPa is acting on the shaded part of the soil shown in Figure 2-33. Calculate the vertical stress at a depth of 3 m below point A.

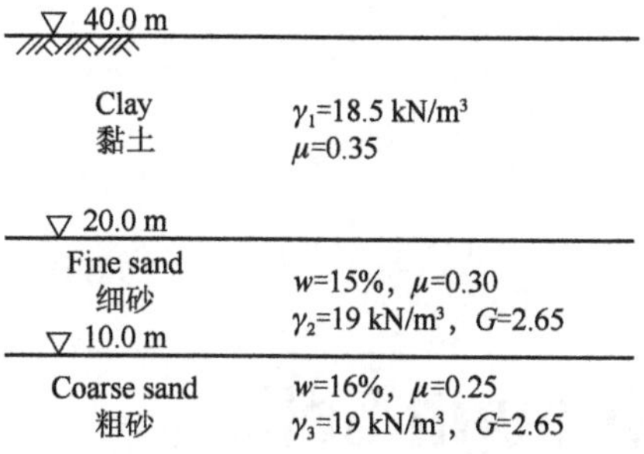

▽ 40.0 m

Clay
黏土
γ_1=18.5 kN/m^3
μ=0.35

▽ 20.0 m

Fine sand
细砂
w=15%, μ=0.30
γ_2=19 kN/m^3, G=2.65

▽ 10.0 m

Coarse sand
粗砂
w=16%, μ=0.25
γ_3=19 kN/m^3, G=2.65

Figure 2-32　Soil layer distribution
图 2-32　土层分布

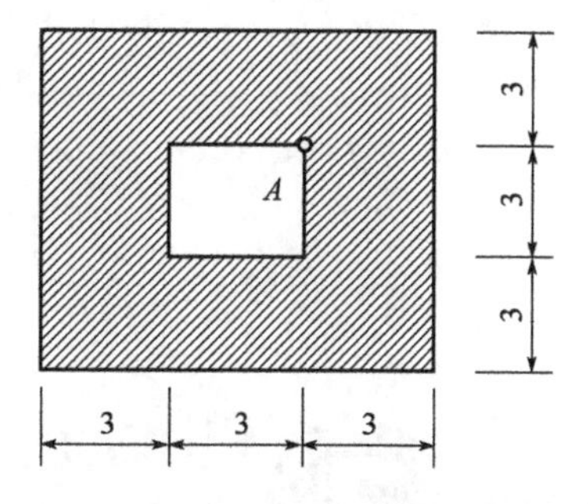

Figure 2-33　The region under uniformly distributed pressure (unit: m)
图 2-33　均布荷载作用范围(单位:m)

(3) The base pressure distribution of a strip foundation is shown in Figure 2-34. Calculate the vertical stress at depths at 2.5 m, 5 m, and 10 m below points A, B, and C.

(3)条形基础的基底压力分布如图 2-34 所示,计算 A、B、C 三点以下 2.5 m、5 m、10 m 处竖向应力。

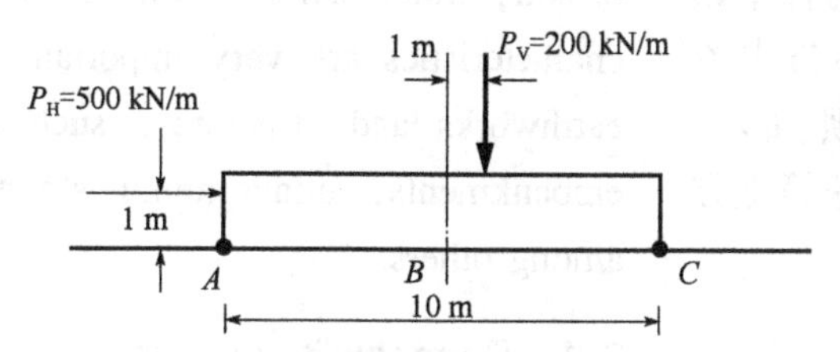

Figure 2-34　Foundation pressure distribution
图 2-34　基底压力分布

(4) A uniformly distributed pressure p = 40 kPa is applied to a circular foundation, as shown in Figure 2-35. Calculate the vertical stress at depths of 2 m, 4 m, 6 m, and 10 m under the midpoint O and edge point A.

(4)圆形基础上作用均布荷载 p = 40 kPa,如图 2-35 所示,求基础中点 O 和边点 A 下 2 m、4 m、6 m和 10 m 深度处的竖向应力。

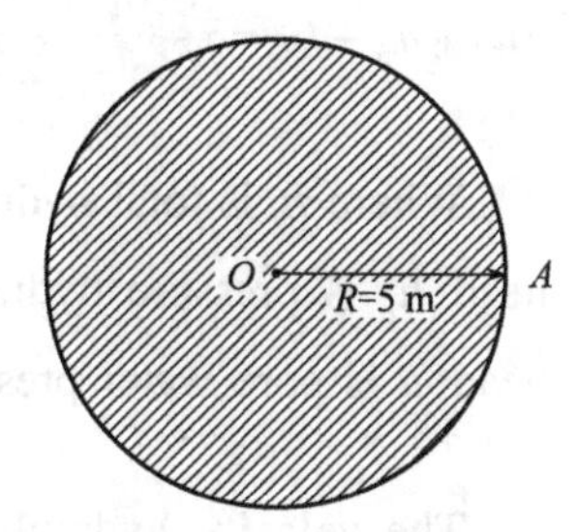

Figure 2-35　The region under uniformly distributed pressure
图 2-35　均布荷载作用范围

Chapter 3 Flow of Water through Soils 土的渗流

水可以在土体中连通的孔隙间流动,在岩土工程的各个领域内,许多工程应用都与水流(渗流)特性相关,如土坝、堤坝、路堤、地下构筑物和开挖等土方工程。

Because of the relatively large proportion of void space of soil, water can flow through it. Water flow (seepage) characteristics are very important in many applications of earthworks and structures, such as earth dams, levees, embankments, underground structures, and excavations, among others.

3.1 土的渗透性

3.1 Permeability of soils

3.1.1 水头

水头指单位重量水体所具有的能量。按照伯努利方程,土体渗流中的总水头计算方法为:

3.1.1 Hydraulic heads

Hydraulic heads refers to the mechanical energy per unit weight. Bernoulli's equation [Equation (3-1)] is used to define the flow of water through soil masses:

$$h_t = h_z + h_p + h_v = z + \frac{u}{\gamma_w} + \frac{v^2}{2g} \tag{3-1}$$

式中,h_t为总水头;h_z为位置水头;h_p为压力水头;h_v为流速水头;u为孔隙水压力;v为渗流流速。

Where h_t is total hydraulic head, h_z is elevation hydraulic head, h_p is pressure hydraulic head, h_v is velocity hydraulic head, u is pore water pressure, v is flow velocity.

土体中渗流速度v一般很小,与其他项相比,流速水头$v^2/2g$一般很小,为简便起见一般忽略。式(3-1)简化为:

The velocity hydraulic head term $v^2/2g$ is neglected in most soil mechanics problems since this value is quite small in comparison with the values of other terms. Thus, Equation (3-1) becomes:

$$h_t = h_z + h_p = z + \frac{u}{\gamma_w} \tag{3-2}$$

Consider a cylinder containing a soil mass with water flowing through it at a constant rate, as depicted in Figure 3-1. Two tubes, A and B (called piezometers) are connected to the cylinder at a distance L apart. As the water flows through the soil, energy is dissipated through friction with the soil particles, resulting in a loss of hydraulic head. The head loss between A and B, assuming a decrease in hydraulic head, is positive, and its value at an arbitrarily selected position at the top of the cylinder is $\Delta H=(h_p)_A-(h_p)_B$.

图 3-1 表示一个装有土体且有水以恒定速率流过的圆柱体试样,在试样两端 A、B 点处设置测压管,距离为 L。由于土颗粒的阻力作用,渗流过程存在能量耗散,导致水头损失。A 点与 B 点之间存在水头损失,假设水头损失为正,在试样顶部任意选择基准时的水头差为 $\Delta H=(h_p)_A-(h_p)_B$。

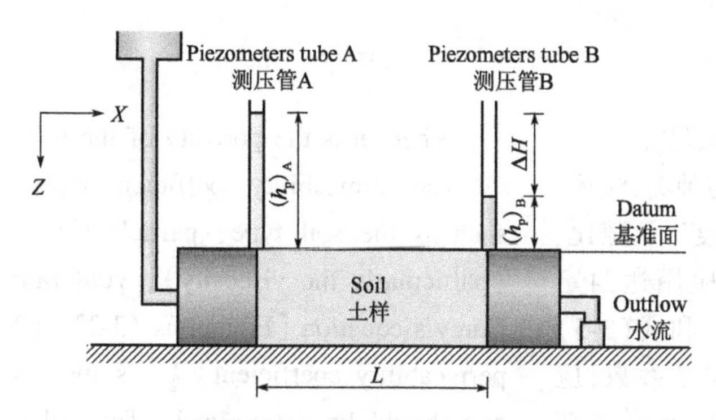

Figure 3-1 Hydraulic head loss due to flow of water through soil

图 3-1 土体渗流引起的水头损失

3.1.2 Darcy's equation

Darcy (1856) proposed that the average flow velocity through soils is proportional to the gradient of the total hydraulic head:

3.1.2 达西定律

达西(1856)指出土体中水渗流的平均速度与总水头的水力坡降成正比:

$$v=ki \tag{3-3}$$

Where v is the discharge velocity of water flow through porous media (m/s); k is the (m/s), and i is the hydraulic gradient (permeability coefficient head loss/flow length $=\Delta h/L$)

式中,v 为多孔介质断面平均渗透速度(m/s);k 为土的渗透系数(m/s);$i(=\Delta h/L)$ 为水力梯度,即两点之间的总水头差与径流长之比。

The seepage discharge, q, is the product of the average velocity, v, and the cross-sectional area, A.

渗水量 q 为平均渗流速度 v 与断面面积 A 之积。

$$q=vA=kiA=k(\Delta h/L)A \tag{3-4}$$

$$Q=qt=kiAt=(k\Delta hAt)/L \tag{3-5}$$

Where q is the units of measurement (m^3/s); Q is the total amount of flow (m^3); t is a time period (s).

式中,q 为单位时间透过土体的水量(m^3/s);Q 为总渗水量(m^3);t 为时间(s)。

需要注意的是,式(3-3)中渗流速度 v 并不是土中水的实际流速,而是一种假想的平均流速,假定水在土体中的渗流是通过整个土体截面来进行的。而实际上,水仅通过土体中的孔隙流动,水在土体中的实际速度大于式(3-3)的计算结果。实际流速 v_s的计算方法为:

Note that the discharge velocity v in Equation (3-3) is not the true velocity of water flow but rather is an average velocity in the flow direction through the porous media. Since water can flow only in the void section of the medium, the true velocity of water (still an average in the direction of the average flow direction) must be faster than v to carry the same quantity of water. The true velocity through the void is called the seepage velocity v_s and is computed as:

$$v_s = \frac{v}{n} \tag{3-6}$$

式中,n 为土的孔隙率。

Where n is the porosity of the soil.

渗透系数与土的种类、粒径、流体性质(特别是黏度)、孔隙比、孔径等因素有关。应用达西定律式(3-3)、式(3-4)和式(3-5)时,渗透系数 k 是关键参数,应予以确定。k 值以指数形式变化。例如,碎石土的 k 值大于 1×10^{-1} cm/s,而黏性土的 k 值小于 1×10^{-7} cm/s。常见土的渗透系数如表 3-1 所示。

The permeability coefficient depends on various factors, such as the soil type, particle size, pore fluid properties (particularly the viscosity), void ratio, and pore size. In Darcy's equation [Equations (3-3), (3-4) and (3-5)], the permeability coefficient, k, is the sole material parameter and should be determined. The value of k changes in an exponential manner. For example, the value of k is more than 1×10^{-1} cm/s for gravels but is less than 1×10^{-7} cm/s for clayey soils. Table 3-1 shows typical ranges of k values for different types of soils.

Table 3-1 Typical k values for different soils

表 3-1 不同类型土的典型渗透系数 k 值

Relative permeability 相对渗透性	Permeability coefficient k (cm/s) 渗透系数	Typical soils 典型土类
Very permeable 强渗透性	$>1\times10^{-1}$	Coarse gravel 粗砾石
Medium permeable 中渗透性	$1\times10^{-1} \sim 1\times10^{-3}$	Sand, fine sand 砂,细砂
Low permeable 低渗透性	$1\times10^{-3} \sim 1\times10^{-5}$	Silty sand, dirty sand 粉砂,淤积砂
Very low permeable 极低渗透性	$1\times10^{-5} \sim 1\times10^{-7}$	Silt, fine sandstone 粉土,细砂岩
Impervious 不可渗透的	$<1\times10^{-7}$	Clay 黏土

Example 3-1

Figure 3-2 shows water flow through a soil specimen in a cylinder. The specimen's k value is 3.4×10^{-4} cm/s. Calculate the pressure head h_p at points A, B, C, and D, and draw the levels of water height in standpipes.

例题 3-1

图3-2 所示为一个水通过圆柱体土样的渗流试验，土样的 k 值为 3.4×10^{-4} cm/s。计算 A、B、C 和 D 点的水头 h_p，并绘制立管中的水位高度。

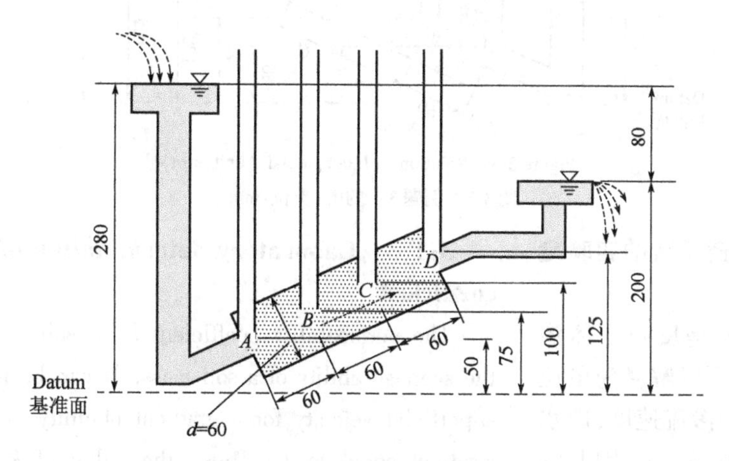

Figure 3-2 Diagram for Example 3-1 (unit: mm)

图 3-2 例题 3-1 的附图(单位:mm)

Solution

Based on the datum selected in the figure and using Equation (3-2), a computation table is constructed in Table 3-2. In the table, for calculating h_t, the hydraulic head loss from A to B is one third of the total hydraulic head loss (80 mm). The same total hydraulic head loss occurs from B to C and from C to D. The heights of water in standpipes are plotted in Figure 3-3.

解：

结合式(3-2)和图中数据，建立计算表 3-2。表中计算 h_t 时，从 A 到 B 的水头损失为总水头损失(80 mm)的三分之一，相同的总水头损失发生在 B 到 C 和 C 到 D 之间，立管中水的高度如图3-3 所示。

Table 3-2 Hydraulic heads, h_z, h_t, and h_p at various points in Figure 3-2

表 3-2 图 3-2 中不同位置的水头 h_z、h_t和 h_p

点 Point	h_z(mm)	h_t(mm)	$h_p = h_t - h_z$(mm)
A	50	280.0	230.0
B	75	280.0 − 80/3 = 253.3	178.3
C	100	253.3 − 80/3 = 226.6	126.6
D	125	226.6 − 80/3 = 200.0	75.0

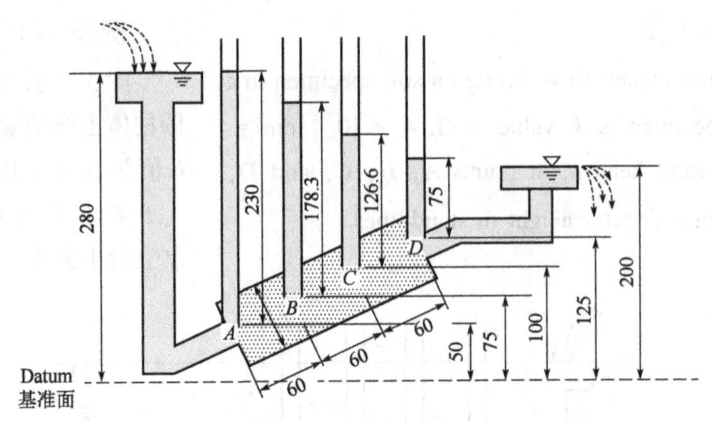

Figure 3-3 Solution to Example 3-1 (unit:mm)
图 3-3 例题 3-1 的解(单位:mm)

3.1.3 渗透系数的实验室测定方法

渗透系数 k 是反映土体渗透能力的参数,可以解释为单位水力梯度对应的渗流速度,即水力梯度等于 1 时,$k=v$。因此,k 值是衡量土体渗透能力的重要指标。

两种常用的渗透系数室内测定方法包括常水头试验和变水头试验。需要注意的是,室内渗透试验测得的渗透系数通常不准确,原因有很多,例如各向异性(即水平流动和垂直流动的 k 值不同),以及小试样无法体现现场大体积土体的真实情况。一般地,现场试验测得的 k 值更为准确。

1)常水头试验

常水头试验适用于透水性强的粗颗粒土。常水头试验装置如图 3-4 所示,土样放置在一个圆柱形模具内,通过调节供水量来保持流经土样的恒定水头损失 h。将流出试样的水收集在量筒中,并记录试验持续时间。根据达西定律,时间 t 内的总流量 $Q=vAt=kiAt$。

3.1.3 Laboratory determination of permeability coefficient

The permeability coefficient k, a soil parameter, reflects the seepage ability of a soil mass. It can be interpreted as the superficial velocity for a gradient of unity, i. e., $k=v$ for a gradient equal to 1. Thus, the value of k is an important measure of a soil's seepage ability.

The two most common laboratory methods for determining the permeability coefficient of a soil are the constant-hydraulic-head test and falling-hydraulic-head test. It should be noted that values of the permeability coefficient measured in laboratory permeameter tests are often highly inaccurate due to various reasons, such as anisotropy (i. e., values of k being different for horizontal and vertical flow) and small samples being unrepresentative of large volumes of soil in the ground. In practice, values of k measured from in situ tests are more reliable than those measured in the laboratory.

1) Constant-hydraulic-head test

The constant-hydraulic-head test is suitable for very permeable granular materials. The basic laboratory test arrangement is shown in Figure 3-4. The soil specimen is placed inside a cylindrical mold, and the constant-hydraulic-head loss h of water flowing through the soil is maintained by adjusting the supply. The outflow water is collected in a measuring cylinder, and the duration of the collection period is recorded. From Darcy's law, the total quantity of flow Q during time t is given by $Q=vAt=kiAt$.

$$k = \frac{QL}{Aht} \tag{3-7}$$

Where A is the cross-sectional area of the specimen; i is h/L; L is the length of the specimen; Q is $k(h/L)At$.

式中，A 为试样横截面积；i 为 h/L；L 为试样长度；Q 为 $k(h/L)At$。

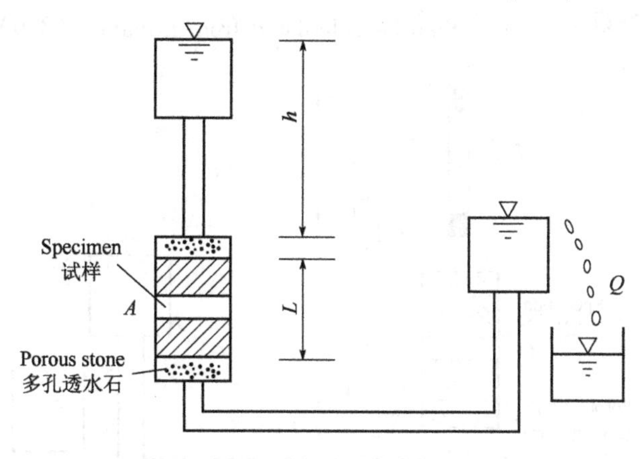

Figure 3-4 Constant-head laboratory permeability test

图 3-4 常水头试验

Once all the quantities on the right-hand side of Equation (3-7) have been determined from the test, the permeability coefficient of the soil can be calculated.

通过试验可得到式(3-7)右侧的所有参数，则通过计算可得到土体的渗透系数。

2) Falling-hydraulic-head test

2)变水头试验

The falling-hydraulic-head permeability test is more suitable for fine-grained soils. Figure 3-5 shows the typical laboratory arrangement for the test. The soil specimen is placed inside a tube, and a standpipe is attached to the top of the specimen. Water that comes from the standpipe flows through the specimen. The initial hydraulic head difference h_1 at time $t = t_1$ is recorded, and then water is allowed to flow through the soil such that the final hydraulic head difference at time $t = t_2$ is h_2.

变水头试验适用于细粒土，变水头试验装置如图 3-5 所示，土样放置在一根细管内，试样的顶端与一根直管连接。从竖直细管流出的水流过土样，记录时间 $t = t_1$ 时的初始水头差 h_1 以及时间 $t = t_2$ 时的最终水头差 h_2。

The rate of flow through the soil is:

流经土样的水量为：

$$q = kiA = k\frac{h}{L}A = -a\frac{\mathrm{d}h}{\mathrm{d}t} \tag{3-8}$$

Where h is the head difference at any time t; A is the cross-sectional area of the specimen; a is the cross-sectional area of the standpipe, and L is the length of the specimen.

式中，h 为不同时刻的水头差；A 为试样横截面积；a 为直管横截面积；L 为试样长度。

From Equation (3-8):

根据式(3-8)有：

$$k=\frac{aL}{A(t_2-t_1)}\ln\frac{h_1}{h_2} \tag{3-9}$$

h_1和h_2分别为t_1和t_2时刻的水头高度;a、L、A、t_1、t_2、h_1和h_2的值可通过试验确定,根据式(3-9)即可计算土体的渗透系数k。

h_1 and h_2 are the hydraulic head height at times t_1 and t_2. The values of a, L, A, t_1, t_2, h_1 and h_2 can be determined from the test, and the permeability coefficient k for a soil can then be calculated from Equation (3-9).

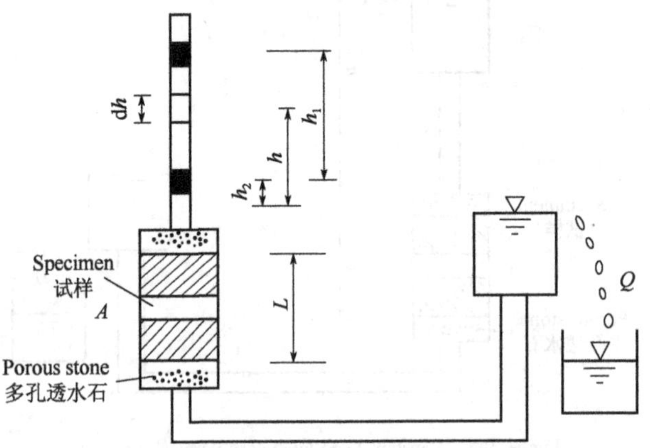

Figure 3-5 Falling-hydraulic-head laboratory permeability test

图 3-5 变水头试验

3.2 渗透力及渗透稳定

3.2.1 渗透力

水在土体中渗流时会受到土颗粒的阻滞作用,从而产生水头损失。反之,水流也对土颗粒产生一个反作用力。单位体积土体受到的渗流作用力称为渗透力,以D表示。

为估算作用在单位体积土体上的渗透力,取一个长度为L、横截面积为A的水柱,如图3-6所示。水柱的重量为$\gamma_w LA$,B点处横截面上的静水压力为$\gamma_w h_1 A$,A点处横截面上的静水压力为$\gamma_w h_2 A$。单位体积土体受到的渗流作用力为T(kN/m^3),根据力的平衡条件可得:

3.2 Seepage force and stability

3.2.1 Seepage force

As water flows through soil, it is blocked by soil particles, resulting in hydraulic head losses and simultaneously exerting drag on the soil particles. The drag exerted on a unit volume of soil is defined as the seepage force, which is a body force and is denoted by D.

To evaluate the seepage force per unit volume of a soil, a water column is considered. The length and cross-sectional area of this water column are L and A, respectively, as shown in Figure 3-6. The weight of the water column is $\gamma_w LA$; the hydrostatic force on side B is $\gamma_w h_1 A$, and the hydrostatic force on side A is $\gamma_w h_2 A$. The resistance T(kN/m^3) to seepage by soil particles is also a body force; for equilibrium, thus:

$$\gamma_w h_1 A+\gamma_w LA\cos\alpha-\gamma_w h_2 A-TAL=0 \tag{3-10}$$

$$\cos\alpha=\frac{z_1-z_2}{L} \tag{3-11}$$

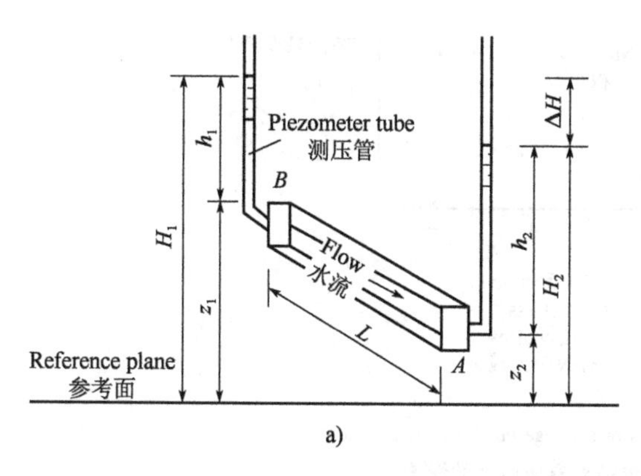

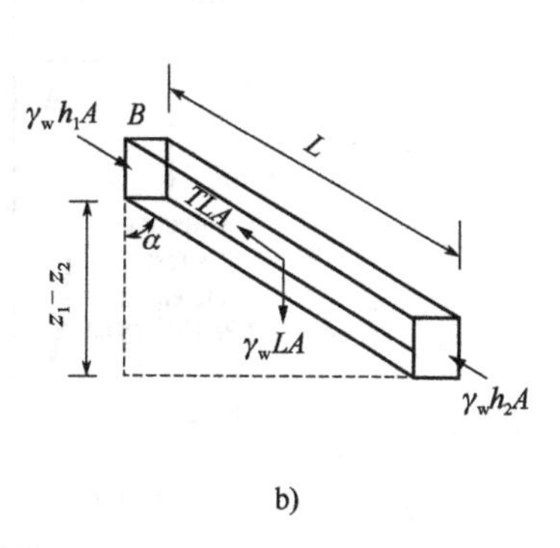

Figure 3-6 Seepage force analysis on the water column

图 3-6 水柱中的渗透力分析

As $z_1+h_1=H_1$, $z_2+h_2=H_2$; therefore, Equations (3-10) and (3-11) yield:

因为 $z_1+h_1=H_1$, $z_2+h_2=H_2$,因此式(3-10)和(3-11)变换为:

$$T=\gamma_w i \tag{3-12}$$

The direction of permeability D is opposite that of T; hence, the value of the seepage force is:

渗透力 D 的方向与 T 相反,其值为

$$D=\gamma_w i \tag{3-13}$$

Based on Equation (3-13), the seepage force D is proportional to the hydraulic gradient i. The direction of the seepage force is the same as the flow direction.

由式(3-13)可知,渗透力 D 与水力梯度 i 成正比,渗透力方向与水的渗流方向一致。

3.2.2 Upward seepage through soil

3.2.2 土体发生向上渗流

With changes in the flow direction, the seepage force will have different effects on soil, as shown in Figure 3-7. When downward seepage occurs, the direction of theseepage force and gravity are the same, increasing the effective stress. When upward seepage occurs, the directions of the seepage force and gravity are opposite to each other, decreasing the effective stress, and the soil weight decreases with increasing pore water pressure.

随着水流方向的变化,渗流力对土体的作用也会不同,如图 3-7所示。当向下渗流时,渗流力的方向与重力方向相同,有效应力增大;当向上渗流时,渗流力和重力方向相反,有效应力减小。土体自重应力随孔隙水压力的增大而减小。

Consider the special case of seepage vertically upward. If the seepage force D exerted on the soil is equal to the effective gravity W' [Equation (3-14)], the soil loses its strength and behaves similar to a viscous fluid. The soil state at which the strength is zero is called static liquefaction.

考虑渗流垂直向上的特殊工况,如果渗透力 D 等于土的有效重量 W'[式(3-14)],则土体失去强度,其表现类似于黏性流体。强度为零的土体状态称为静态液化。

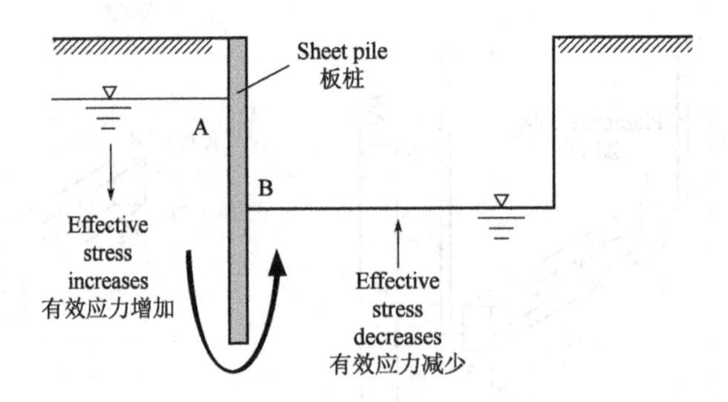

Figure 3-7 Effects of seepage on the effective stress

图3-7 渗流对有效应力的影响

$$D = \gamma_w i = \gamma' \tag{3-14}$$

如果垂直向上的渗透力超过土体浮重度,土颗粒会被水流携带向上运移并沉积在地表,称为流土现象。

If the upward seepage forces exceed the buoyant weight, the particles may be carried upward to be deposited at the ground surface, which is called flowing soil.

与合力为0相对应的水力梯度值称为临界水力梯度(i_{cr})。

The value of the hydraulic gradient corresponding to a zero resultant body force is called the critical hydraulic gradient (i_{cr}).

$$i_{cr} = \frac{\gamma'}{\gamma_w} = \left(\frac{G_s - 1}{1 + e}\right)\frac{\gamma_w}{\gamma_w} = \frac{G_s - 1}{1 + e} \tag{3-15}$$

式中,G_s为土粒相对密度;e为孔隙比。

Where G_s is the specific gravity and e is the void ratio.

由于G_s是常数,土体的临界水力梯度仅是孔隙比的函数。为使设计的结构处于稳定状态,应保证水力坡降不超过临界水力梯度。

Since G_s is constant, the critical hydraulic gradient is solely a function of the void ratio of the soil. When designing structures that are subjected to steady-state conditions, it is essential to ensure that no critical hydraulic gradient will develop.

例题 3-2

如图3-8所示,坑底以下黏土层厚度为5 m,黏土层下有承压水,用测压管测量水头。在施工过程中,基坑降水使地下水位保持在基坑底部0.5 m以下。黏土重度(γ)为17 kN/m^3,饱和重度(γ_{sat})为18.6 kN/m^3,判断坑底是否会隆起。

Example 3-2

As shown in Figure 3-8, the thickness of the clay layer under a pit bottom is 5 m. There is confined water under the clay layer, and the hydraulic pressure height is measured by a piezometer tube. During construction, the groundwater level is maintained at a depth 0.5 m below the pit bottom by foundation pit dewatering. The unit weight (γ) of the clay is 17 kN/m^3, and its saturated unit weight (γ_{sat}) is 18.6 kN/m^3. Determine whether a pit bottom upheaval can occur.

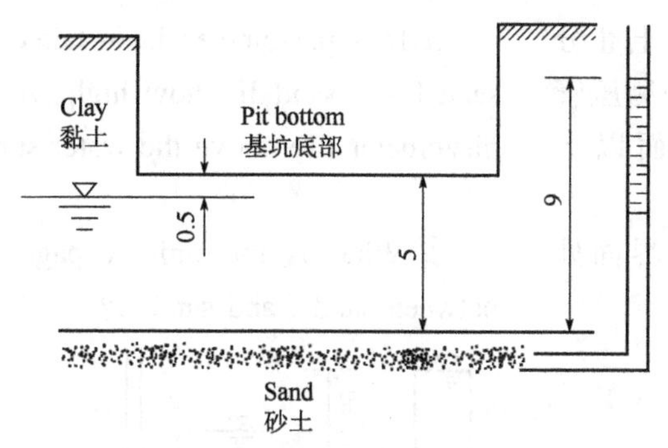

Figure 3-8 Diagram for Example 3-2 (unit: m)

图 3-8 例题 3-2 的附图(单位:m)

Solution 1

Since confined water exists, an upward seepage force will be exerted, which is given by:

$$D_{\mathrm{d}} = \gamma_{\mathrm{w}} \cdot i = \gamma_{\mathrm{w}} \cdot \frac{\Delta h}{L} = 9.8 \times \frac{4.5}{4.5} = 9.8(\mathrm{kN/m^3})$$

Through the 4.5 m flow path, the seepage force is expressed by the surface force.

$$D'_{\mathrm{d}} = D_{\mathrm{d}} \cdot L = 9.8 \times 4.5 = 44.1(\mathrm{kPa})$$

The effective weight of the soil layer above and below the groundwater is given by:

$$W' = 17 \times 0.5 + (18.6 - 9.8) \times 4.5 = 48.1(\mathrm{kPa})$$

Since $D_{\mathrm{d}} \leqslant W'$, pit bottom upheaval cannot occur.

Solution 2

The pressure hydraulic head difference of the confined water is $H = 9 - 4.5 = 4.5$ m. Hence:

$$\gamma_{\mathrm{w}} H = 9.8 \times 4.5 = 44.1(\mathrm{kPa})$$

The effective weight of the soil layer between the pit bottom and the top surface of the confined water is given by:

$$\Sigma \gamma_i h_i = 17 \times 0.5 + (18.6 - 9.8) \times 4.5 = 48.1(\mathrm{kPa})$$

Since $\Sigma \gamma_i h_i > \gamma_{\mathrm{w}} H$, pit bottom upheaval cannot occur.

解 1:

由于承压水的存在,会产生向上的渗透力,即:

通过 4.5 m 长的渗流路径,渗流力用面力表示,即

地下水位上下土层的有效重量为:

由于 $D_{\mathrm{d}} \leqslant W'$,坑底不会隆起。

解 2:

承压水的水头差为 $H = 9 - 4.5 = 4.5$ m,因此:

坑底至承压水顶面之间土层的有效重量为:

由于 $\Sigma \gamma_i h_i > \gamma_{\mathrm{w}} H$,坑底不会隆起。

Exercises

(1) A permeation device is shown in Figure 3-9. The permeability coefficient of sand Ⅰ is $k_1 = 2 \times 10^{-1}$ cm/s, the permeability coefficient of sand Ⅱ is $k_2 = 1 \times 10^{-1}$ cm/s, and the cross-sectional area of sand sample $A = 200$ cm^2.

习 题

(1)某渗透装置如图 3-9 所示,砂土Ⅰ的渗透系数 $k_1 = 2 \times 10^{-1}$ cm/s,砂土Ⅱ的渗透系数 $k_2 = 1 \times 10^{-1}$ cm/s,砂土样横截面积 $A = 200$ cm^2,试问:

①若在砂土Ⅰ与砂土Ⅱ分界面处安装一测压管,则测压管中水面将升至右端水面以上多高?

②砂土Ⅰ与砂土Ⅱ界面处的单位渗水量 Q 多大?

①If a piezometer is installed at the interface between sand I and sand II, how high will the water surface in the piezometer rise above the water surface at the right end?

② What is the unit seepage flow Q at the interface between sand I and sand II?

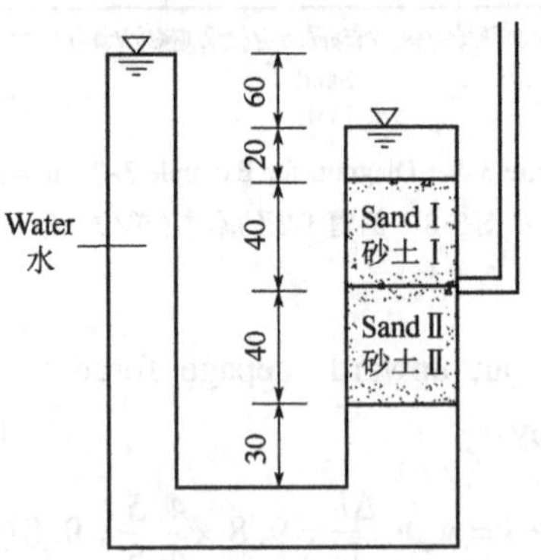

Figure 3-9 Diagram for Exercise (1) (unit:cm)

图3-9 习题(1)的附图(单位:cm)

(2)常水头渗透试验中,已知渗透仪直径 $D=75$ mm,在 $L=200$ mm渗流路径上的水头损失 $h=83$ mm,在60s时间内的渗水量 $Q=71.6$ cm^3,求土的渗透系数。

(2)In a constant-head permeability test, the diameter of the permeameter is $D=75$ mm; the hydraulic head loss is $h=83$ mm; the distance of the permeameter is $L=200$ mm, and the amount of seepage in 60 s is $Q=71.6$ cm^3. Calculate the permeability coefficient of the soil.

(3)如图3-10所示,在长为10 cm,横截面积为8 cm^{2}的圆筒内装满砂土,管内水位高出筒5 cm(固定不变),流水自下而上通过试样后可溢流出去。①计算渗透力的大小,判断是否会产生流土现象;②计算临界水力梯度值。

(3)As shown in Figure 3-10, a cylinder with a length of 10 cm and an area of 8 cm^2 is filled with sand. The water level in the tube is 5 cm higher than that in the tube (fixed), and the flowing water can overflow after passing through the sample from bottom to top. ①Calculate the seepage force and judge whether there will be a flowing soil phenomenon, and ②calculate the critical hydraulic gradient value.

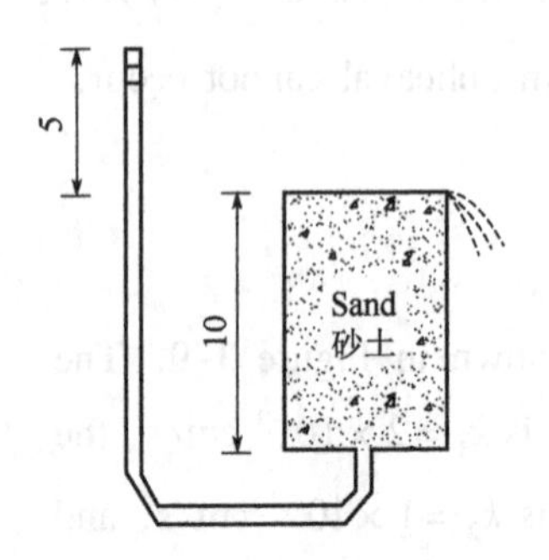

Figure 3-10 Diagram for Exercise (3) (unit:cm)

图3-10 习题(3)的附图(单位:cm)

(4) As shown in Figure 3-11, under a constant total head difference, water flows through two soil samples from bottom to top and overflows from the top of soil specimen 1.

①Taking the bottom surface $C-C$ of soil specimen 2 as the datum plane, calculate the total hydraulic head and elevation hydraulic head of the surface.

②Given that the head loss of water flowing through soil specimen 2 is 30% of the total hydraulic head difference, calculate the total hydraulic head and elevation hydraulic head of the $B-B$ plane.

③Given that the permeability coefficient of soil specimen 2 is 0.05 cm/s, calculate the discharge per unit area of the cross section of the soil specimen per unit time.

④Calculate the permeability coefficient of soil specimen 1.

(4)如图 3-11 所示,在恒定的总水头差之下,水自下而上透过两个土样,从土样 1 顶面溢出。

①已知土样 2 底面 $C—C$ 为基准面,求该面的总水头和位置水头;

②已知水流经土样 2 的水头损失为总水头差的 30%,求 $B—B$ 面的总水头和位置水头;

③已知土样 2 的渗透系数为 0.05 cm/s,求单位时间内通过土样横截面单位面积的流量;

④求土样 1 的渗透系数。

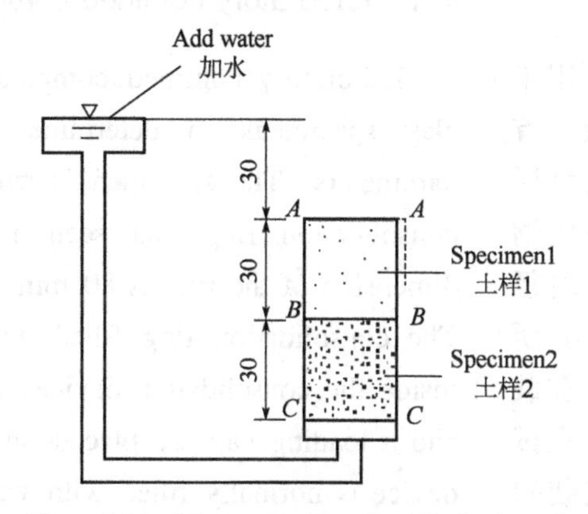

Figure 3-11 Diagram for Exercise (4) (unit:cm)

图 3-11 习题(4)的附图(单位:cm)

Chapter 4 Consolidation and Settlement of Soils 土的固结与沉降

4.1 室内侧限压缩试验

室内侧限压缩试验,常用于测定黏性土的压缩性参数。首先将试样置入金属环刀内,然后将土样连同环刀装入侧限压缩仪内,环刀的尺寸一般为内径80 mm、高20 mm,如图4-1所示,土样顶部放置透水石和加载帽。水槽内充水,以防土样在试验过程中干燥。通过加载装置给土样施加均匀固结压力,并通过一个百分表测读试件的竖向变形。在固结试验过程中,随着土样中水的排出,土样发生竖向压缩变形。

4.1 Laboratory confined compression test

Laboratory confined compression test is performed for clay specimens to determine several key consolidation parameters. The specimen is trimmed to fit inside a rigid consolidation ring, as seen in Figure 4-1. A typical dimension of the ring is 80 mm diameter and 20 mm high. The consolidation ring filled with the specimen is placed inside the consolidation device, and the upper porous stone and a loading cap are placed on top of the specimen. The device is normally filled with water to avoid drying out of the specimen during the test. A consolidation load is applied and the vertical deformation is monitored with a dial gauge. In this system, soil deforms only in a vertical direction due to the escape of water during the consolidation process.

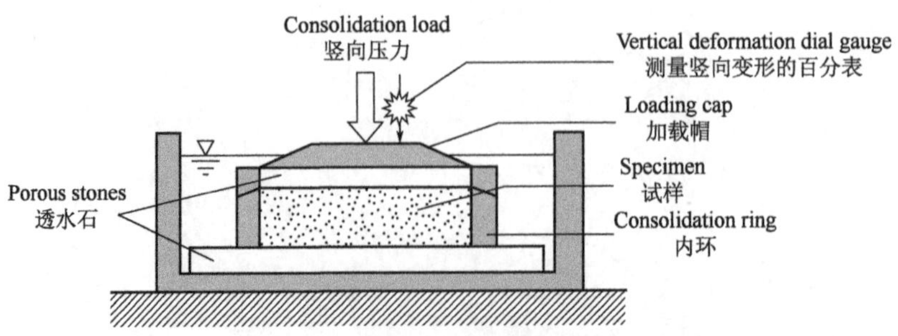

Figure 4-1 Confined compression test setup

图4-1 侧限压缩试验装置

将各级固结应力(通常为50 kPa、100 kPa、200 kPa、300 kPa

Laboratory data are analyzed for the final settlements achieved under given consolidation stresses σ (typical values of 50 kPa, 100 kPa, 200 kPa, 300 kPa, 400 kPa) at the end of consolidations. Relationships between consolidation stress σ (in log scale) and final void ratio e are plotted in Figure 4-2. This curve is called the e-logσ curve. The loading curve (decreasing e with increase in σ) and unloading curve (increasing e with decease in σ) are seen in the figure.

和 400 kPa)及对应的土样最终变形量绘制在半对数坐标上，便得到土体压缩曲线，称为 e-logσ 曲线，加载曲线(e 减小、σ 增大)和卸载曲线(e 增大、σ 减小)如图 4-2 所示。

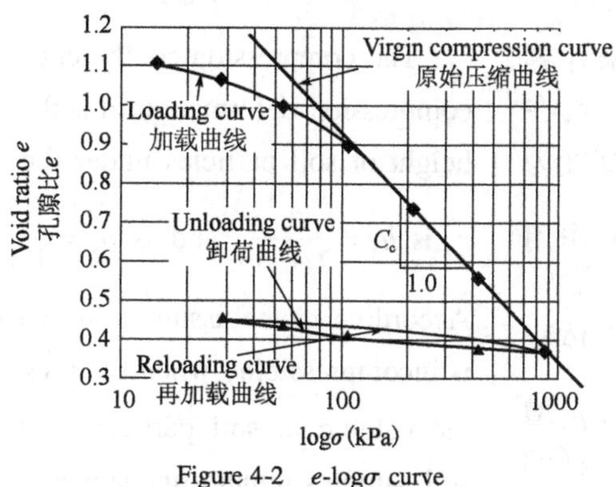

Figure 4-2 e-logσ curve

图 4-2 e-logσ 曲线

1) Compression coefficient a

The degree of steep drop of the compression curve represents the compressibility of the soil. Therefore, the tangent slope a at any point of the curve indicates the compressibility under the corresponding pressure p, as shown in Figure 4-3, namely:

1)压缩系数 a

压缩曲线的变化程度体现了土的压缩特性，曲线上任一点的切线斜率 a 表示对应压力 p 作用下土的压缩性，如图 4-3 所示，即：

$$a = -\frac{\mathrm{d}e}{\mathrm{d}p} \tag{4-1}$$

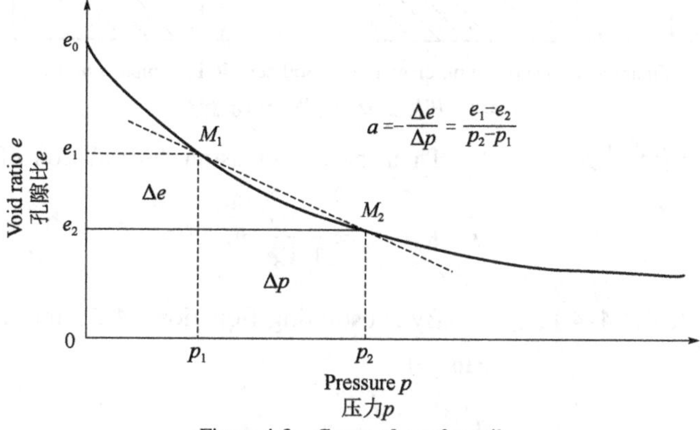

Figure 4-3 Curve of e-p for soil

图 4-3 土的 e-p 曲线

Where negative sign means that e decreases with the increase of p. In practice, a point in general soil increases from the original self weight stress p_1 to the soil stress p_2 under the action of external load. When the pressure range

式中，负号表示随着压力 p 的增加，孔隙比 e 逐渐减小。实际上，当土中某点受到外荷载作用后，初始自重应力 p_1 会增加到

应力 p_2。当压力变化范围 p_1 ~ p_2不大时,可用线段 M_1M_2代替相应压缩曲线段,此时压缩系数可用割线 M_1M_2斜率表示,则:

p_1-p_2 is small, the corresponding section of M_1M_2 on the compression curve can be replaced by a straight line. In this case, the compressibility can be represented by the slope of the secant M_1M_2 in the graph, hence:

$$a = -\frac{\Delta e}{\Delta p} = \frac{e_1 - e_2}{p_2 - p_1} \tag{4-2}$$

压缩系数 a 可以用来计算土的压缩量。如图 4-4 所示,在自重应力 p_1作用下,土样高度为 $h_s = \frac{h_1}{1 + e_1}$;在自重应力 p_2 作用下,$h_s = \frac{h_2}{1 + e_2}$。根据土颗粒体积不可压缩的假设,当压力由 p_1增加到 p_2时,土颗粒体积和土样的横截面积不变,则:

The compression coefficient a can be used to calculate the compression deformation of soil. As shown in Figure 4-4, the height of soil particles under the action of self weight stress p_1 is $h_s = \frac{h_1}{1 + e_1}$, and is $h_s = \frac{h_2}{1 + e_2}$ under the action of p_2. According to the assumption that the volume of soil particles is incompressible, when the pressure increases from p_1 to p_2, the volume of soil particles and the cross-sectional area of soil samples remain unchanged, hence:

$$\frac{h_1}{1 + e_1} \cdot A = \frac{h_2}{1 + e_2} A \tag{4-3}$$

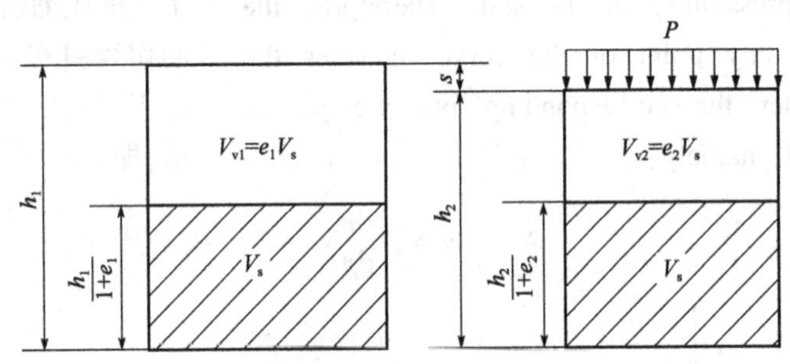

Figure 4-4 Deformation diagram of a soil sample in compression test

图 4-4 压缩试验中土样变形示意图

因此,试样压缩变形量为:

Then the compression deformation of soil is:

$$s = h_1 - h_2 = \frac{e_1 - e_2}{1 + e_1} h_1 \tag{4-4}$$

将式(4-2)代入式(4-4),可得:

By substituting Equation (4-2) into Equation (4-4), we can get:

$$s = \frac{a}{1 + e_1}(p_2 - p_1)h_1 = \frac{a}{1 + e_1}\Delta p h_1 \tag{4-5}$$

式中,Δp 为外荷载引起的附加应力;h_1为土层厚度。

Where Δp is the additional stress caused by external load; h_1 is the thickness of soil layer.

2)压缩指数 C_c

2) Compression index C_c

根据图 4-2 的加载曲线可知,

In the loading section in Figure 4-2, a linear relation is

observed at higher stress level and a straight line is drawn as a virgin compression curve. The slope of the virgin curve is read as compression index C_c and is given by:

在压力较大的部分,e-logσ 近似线性关系,可用直线代替原压缩曲线,原始曲线的斜率称为土的压缩指数 C_c:

$$C_c = \frac{-(e - e_i)}{\log\sigma - \log\sigma_i} = \frac{-(e - e_i)}{\log \dfrac{\sigma}{\sigma_i}}$$

$$\Delta e = e_i - e = C_c \log \frac{\sigma}{\sigma_i} \tag{4-6}$$

Where (e_i, σ_i) and (e, σ) are arbitrary points on the virgin compression curve.

式中,(e_i, σ_i)和(e, σ)是原压缩曲线上的任意点。

Equation (4-6) is used to estimate consolidation settlement for soils that follow only the virgin curve relation. The e-logσ curve is a key relationship to determine final consolidation settlement.

式(4-6)用于计算原压缩曲线关系的土体固结沉降。e-logσ 曲线对建筑物的地基沉降计算具有重要实际意义。

The average slope of the unloading curve and reloading curve in Figure 4-2 is read as compression index C_e, and the compression index C_e does not change with the change of pressure p.

图 4-2 中卸载曲线和再加载曲线的平均斜率称为土的回弹指数或再压缩指数 C_e,C_e不随压力 p 的变化而变化。

3) Oedometric modulus E_s

3)侧限压缩模量 E_s

The ratio of the compressive stress σ_z to the vertical strain ε_z caused by the compressive stress is called the oedometric modulus of the soil, expressed as E_s, i.e:

土在侧限条件下压缩时,受到的压应力 σ_z 与该压应力引起的竖向应变 ε_z 之比称为土的压缩模量,以 E_s表示,即:

$$E_s = \frac{\sigma_z}{\varepsilon_z} \tag{4-7}$$

According to the results of confining compression test, as shown in Figure 4-4, it can be seen that:

如图 4-4 所示,根据侧限压缩试验结果可知:

$$\sigma_z = p_2 - p_1 \tag{4-8}$$

$$\varepsilon_z = \frac{h_1 - h_2}{h_1} = \frac{e_1 - e_2}{1 + e_1} \tag{4-9}$$

Thus:

由此可知:

$$E_s = \frac{p_2 - p_1}{e_1 - e_2}(1 + e_1) = \frac{1}{a}(1 + e_1) \tag{4-10}$$

The reciprocal of the oedometric modulus E_s is defined as the volume compression coefficient m_v of the soil, expressed as $m_v = 1/E_s$.

侧限压缩模量 E_s的倒数称为土的体积压缩系数 m_v,即 $m_v = 1/E_s$。

According to Equation (4-10), the oedometric modulus E_s is a function of the compression coefficient a, and E_s is inversely proportional to a. The larger E_s is, the smaller a

由式(4-10)可知,压缩模量 E_s是压缩系数 a 的函数,E_s与 a 成反比,E_s越大,a 就越小,即土的压缩性越小。当 $E_s<4$ MPa 时,为高压缩性土;当 4 MPa$\leqslant E_s\leqslant$20 MPa 时,为中等压缩性土;$E_s>$20 MPa,为低压缩性土。

is, indicating the smaller the compressibility of the soil. In practice, when $E_s<$ 4 MPa, it is high compressibility soil; when 4 MPa $\leqslant E_s\leqslant$ 20 MPa, it is medium compressibility soil; when $E_s>$ 20 MPa, it is low compressibility soil.

4)变形模量 E_0

4) Modulus of deformation E_0

土在没有侧限(允许侧胀)条件下压缩时,竖向压应力与竖向压应变之比,称为土的变形模量,以 E_0表示:

When the soil is compressed under the condition of no lateral limit (allowable lateral expansion), the ratio of vertical compressive stress to vertical strain is called the deformation modulus of soil, expressed as E_0:

$$E_0=\frac{\sigma_z}{\varepsilon_z} \tag{4-11}$$

土的变形模量在概念上与一般弹性材料的弹性模量是一致的,但在性质上又有区别。线弹性材料在弹性范围内,应力与应变是线性关系,因此弹性模量是一个常量。而土并非理想的弹性体,其变形既有弹性变形又有塑性变形,应力与应变的关系是非线性的,因此土的变形模量是一个变量为了区别,在土力学中将其称为变形模量。

The deformation modulus of soil is consistent with that of general elastic materials in concept. For linear elastic material in the elastic limit, the relationship between the stress and strain is linear, so the elastic modulus is a constant. However, soil is not an ideal elastic body, and its deformation includes both elastic deformation and plastic deformation, hence the relationship between stress and strain is nonlinear, so the deformation modulus of soil is a variable, it is called deformation modulus in soil mechanics.

E_0可结合现场加载试验和弹性力学理论求得,也可根据压缩模量 E_s换算。

E_0 can be obtained by using the elastic mechanics theory according to the field loading test data, and can also be converted according to the compression modulus E_s.

由式(4-11)可知竖向压应变为:

According to Equation (4-11), the vertical compressive strain is:

$$\varepsilon_z=\frac{\sigma_z}{E_s} \tag{4-12}$$

在压缩试验中,土样受三向应力作用,考虑侧向应力 σ_x、σ_y对竖向 ε_z的影响时,根据广义胡克定律得:

In the compression test, when the soil sample is subjected to three-dimensional stress, the influence of the lateral stress σ_x and σ_y on the vertical ε_z is considered, according to the Hooke's law:

$$\varepsilon_z=\frac{1}{E_0}[\sigma_z-\mu(\sigma_x+\sigma_y)] \tag{4-13}$$

由于 $\sigma_x=\sigma_y$,上式可写成

Since $\sigma_x=\sigma_y$, the above equation can be written as follows:

$$\varepsilon_z=\frac{1}{E_0}(\sigma_z-2\mu\sigma_x) \tag{4-14}$$

Substituting $\sigma_x = K_0\sigma_z$, $K_0 = \frac{\mu}{1-\mu}$ into the above equation, we can get:

将 $\sigma_x = K_0\sigma_z$, $K_0 = \frac{\mu}{1-\mu}$ 代入上式得:

$$\varepsilon_z = \frac{\sigma_z}{E_0}\left(1 - \frac{2\mu^2}{1-\mu}\right) \tag{4-15}$$

According to Equations (4-12) and (4-15), the following results can be obtained:

联立式(4-12)与式(4-15),可得:

$$E_0 = \left(1 - \frac{2\mu^2}{1-\mu}\right)E_s = \beta E_s \tag{4-16}$$

Equation (4-16) is only the theoretical relationship between the deformation modulus E_0 and the oedometric modulus E_s. Since the soil is not an ideal elastic body, and the soil samples used in the test are disturbed in the process of sampling, transportation and test operation, it is difficult to maintain its natural structure, so the relationship is approximate. Generally, E_0 value may be several times of E_s, the harder the soil is, the greater the multiple is, while the E_0 value of soft clay is close to E_s value.

式(4-16)只是变形模量 E_0 和压缩模量 E_s 之间的理论关系。由于土不是理想弹性体,加上试验所用土样在取样、运输、试验操作过程中不可避免会受到扰动,很难保持其天然结构。因而,上式只能近似描述变形模量和压缩模量之间的关系。根据统计资料,现场测得的 E_0 值可能是室内侧限压缩试验的 E_s 值的几倍,土愈坚硬则倍数愈大,但是软黏土的 E_0 值与 E_s 值比较接近。

4.2 Normally consolidated and over-consolidated soils

4.2 正常固结土和超固结土

The pre-consolidation stress is the stress at which soil has previously experienced the historical maximum consolidation in the field.

When pre-consolidation stress is found to be the same as the current effective overburden stress σ_0' at the site from which the sample is obtained, the soil is called normally consolidated. Referring to Figure 4-5, a soil has been consolidated at the site under its own weight till sampling takes place (point A). During the sampling process (A to B), the in-situ stress (at A) is reduced to nearly zero (at B) and reloaded in the laboratory consolidation process (B, C to D). The loading path B to C is a reloading process and thus the slope is rather small. After passing the pre-consolidation

土的前期固结压力指土层在历史上曾经承受过的最大有效压力。

当先期固结压力与原状土体上覆有效自重压力 σ_0' 相同时,称为正常固结土。如图 4-5 所示,原状土体在自重作用下固结(A 点);现场取样过程中(A 点至 B 点),应力(A 点)释放至接近 0(B 点);然后在室内固结试验过程(B、C 至 D)中重新加载,加载路径 B 点到 C 点是一个重新加载的过程,其斜率很小;

超过 C 点先期固结压力后,土体进入一个新的应力状态,斜率变陡(斜率接近压缩指数 C_c);达到最大应力 D 点后,卸载过程($D \sim E$)的斜率与 $A \sim B$ 曲线斜率相近。

stress Point C, soil enters a new stress territory and the slope becomes steeper (to the value of C_c). After the maximum stress Point D in the laboratory, the unloading process (from D to E) takes place and its slope is similar to the one of the A to B curve.

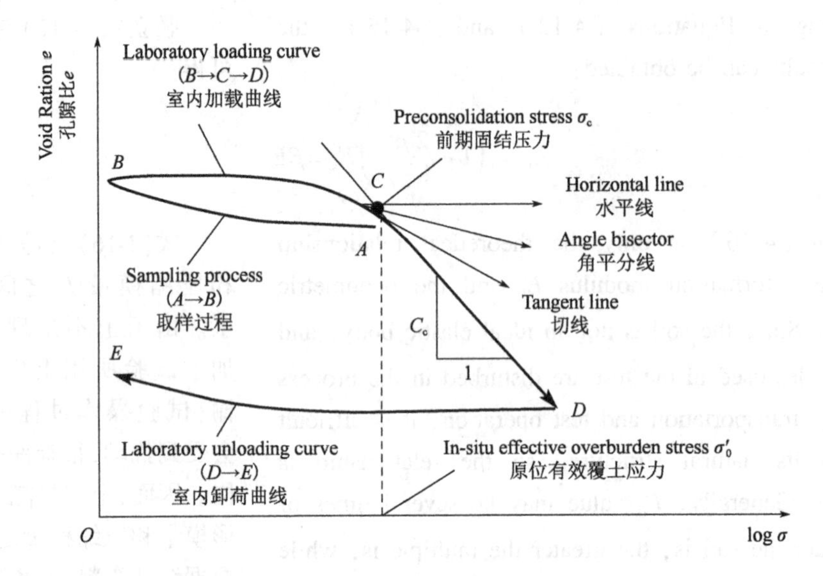

Figure 4-5 e-logσ curve for normally consolidated soil

图 4-5 正常固结土的 e-logσ 曲线

多数情况下,土层的前期固结压力可能高于现有上覆土层的自重压力,这种土称为超固结土。如图 4-6 所示,e-logσ 曲线从 O 点开始,O 点为前期固结压力 $\sigma_{0,\max}$(历史最大有效上覆压力),当前土体自重应力减小到 A 点(当前 σ_0')。值得注意的是,历史最大自重应力 $\sigma_{0,\max}$ 是未知的。

On many occasions, the site might have been subjected to stress higher than σ_0' during its geological history, this soil is called over-consolidated soil. Referring to Figure 4-6, the e-logσ curve starts from Point O with the consolidation stress with $\sigma_{0,\max}$ (historical maximum effective overburden stress), and a portion of the stress has been reduced to Point A (current σ_0'). Note that its historical maximum $\sigma_{0,\max}$ stress cannot be seen visually at the present time.

同理,路径 AB 为取样过程,路径 BCD 为实验室应力加载过程为。e-logσ 曲线上 C 点为先期固结压力,与 $\sigma_{0,\max}$ 值近似。由于冰川融化、开挖、地表侵蚀和地下水位上升等原因,压力 $\sigma_{0,\max}$ 释放到 σ_0。

Similarly, the sampling process follows A to B and the laboratory loading process goes from B, C, and D. Preconsolidation stress found on the laboratory e-logσ curve at Point C is similar to the value of the historical maximum effective overburden stress $\sigma_{0,\max}$. The removal of consolidation stress from $\sigma_{0,\max}$ to σ_0 is due to meltdown of glacial ice, excavation, erosion of top soils, permanent rise of ground water tables, etc.

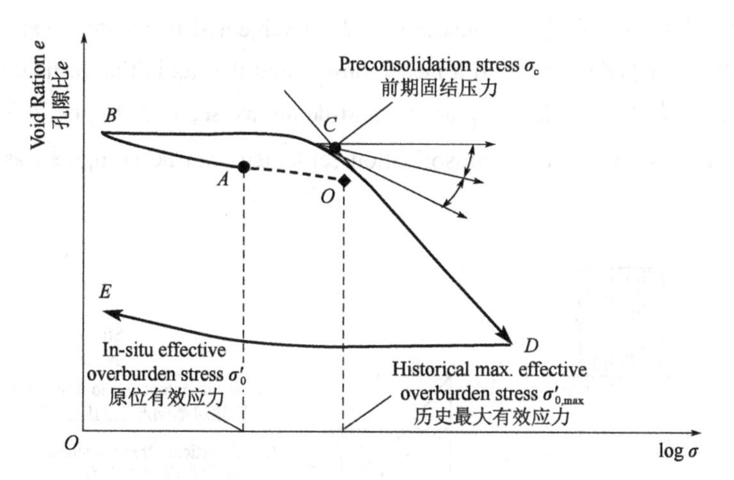

Figure 4-6 e-logσ curve for overconsolidated soil

图4-6 超固结土的 e-logσ 曲线

Over-consolidation ratio (OCR) is defined as the ratio of historical maximum effective overburden stress to current effective overburden stress:

超固结比为土体历史上受的最大有效应力与现在有效应力的比值:

$$\mathrm{OCR} = \frac{\sigma_{0,\max}}{\sigma_0} \tag{4-17}$$

The OCR value for normally consolidated soils is 1.0, and it is higher than 1.0 for overconsolidated soils.

OCR = 1.0 为正常固结土,OCR > 1.0 为超固结土。

Example 4-1

In a city, the area had been covered with a 100 m thick ice load in an early historical time. Some soil in the city is obtained from 10 m deep below the ground surface. The water table was near the ground surface. Estimate the value of OCR for this soil specimen(assume that soil's unit weight is 19 kN/m^3).

例题 4-1

某城市在早期地质历史上曾被100 m厚的冰层所覆盖,现在从这个城市地表以下10 m深处获取部分土体,地下水位接近地表,估算该土体的OCR值(假设土体重度为19 kN/m^3)。

Solution

The ice's unit weight is the same as that of water (9.81 kN/m^3), thus:

解:

冰的重度与水相等(9.81kN/m^3),有:

$$\sigma'_{0,\max} = 9.81 \times 100 + (19 - 9.81) \times 10 = 981 + 91.9 = 1073(\mathrm{kPa})$$

$$\sigma'_0 = (19 - 9.81) \times 10 = 91.9(\mathrm{kPa})$$

$$\mathrm{OCR} = \sigma'_{0,\max}/\sigma'_0 = 1073/91.9 = 11.7$$

4.3 Calculation of consolidation settlement

4.3 地基沉降计算

4.3.1 Final consolidation settlement for thin clay layer

4.3.1 薄土层的最终固结沉降

Assume that a relatively thin clay layer with total

假设厚度为 H 的薄黏土层

由于新建基础而受到一个应力增量 $\Delta\sigma$ 的作用,中间深度处的初始竖向有效应力为 σ_0,如图4-7所示,则最终固结沉降量计算方法如下。

thickness of H is subjected to an incremental stress $\Delta\sigma$ due to a new footing, and that its initial vertical effective stress is σ_0 at its mid-depth as seen in Figure 4-7. Final primary consolidation settlement can be computed as follows.

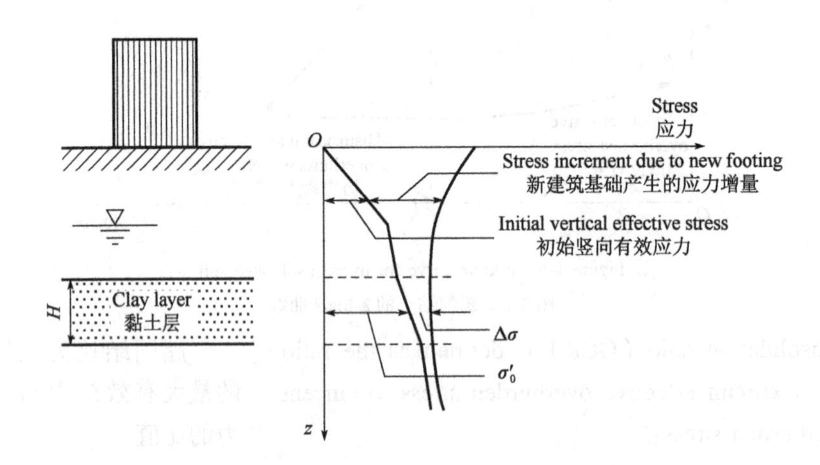

Figure 4-7 Consolidation settlement computation for a thin single clay layer

图4-7 单层薄黏性土的固结沉降计算

1)正常固结土

如图4-8所示,原始压缩曲线上 σ_0 到 $\sigma_0+\Delta\sigma$ 的斜率为 C_c。由式(4-18)得:

1) Normally consolidated soils

As shown in Figure 4-8, σ_0 and $\sigma_0+\Delta\sigma$ are on the virgin curve, and its slope is C_c. In this case, Equation (4-18) is used to calculate Δe as:

$$\Delta e = e_i - e = C_c \log\frac{\sigma_0+\Delta\sigma}{\sigma_0} \tag{4-18}$$

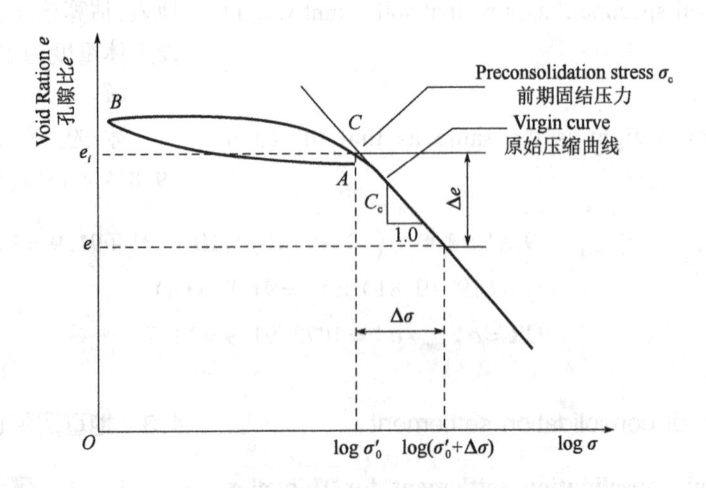

Figure 4-8 Settlement computation for normally consolidated soils.

图4-8 正常固结土的沉降计算

The void ratio change Δe occurs to the total initial height $(1+e_0)$. Thus, proportionally, the final settlement s_f to the total initial clay thickness H is:

初始高度$(1+e_0)$的孔隙比变化量为Δe，最终沉降量s_f与土层初始厚度H的比例关系为：

$$\frac{e}{1+e_0}=\frac{s_f}{H}$$

$$s_f=\frac{H}{1+e_0}e=\frac{H}{1+e_0}C_c\log\frac{\sigma_0+\sigma}{\sigma_0} \tag{4-19}$$

The Δe value can be directly read from the e-logσ curve, and it is applied to the first term of Equation (4-19) to obtain final total consolidation settlement s_f.

也可在e-logσ曲线上直接读取Δe值，并将其代入式(4-19)的第一项，可得最终总固结沉降量s_f。

2) Over-consolidated soils

2)超固结土

For these types of soils, σ_0 and $\sigma_0+\sigma$ are not necessarily on the virgin curve, as seen in Figure 4-9, and thus the constant C_c value with Equation (4-19) cannot be used for the settlement computation. In this case, the Δe value is directly read from the e-logσ curve and substituted into the first term of Equation (4-19):

对于超固结土，σ_0和$\sigma_0+\sigma$不一定在原始压缩曲线上，如图4-9所示，因此式(4-19)中的常数C_c值不能用于沉降计算。在这种情况下，Δe值只能从e-logσ曲线中读取，并代入式(4-19)的第一项：

$$s_f=\frac{H}{1+e_0}e \tag{4-20}$$

It should be noted that for the amount of settlement computation in Equations (4-19) and (4-20), "H" here is always the total thickness of clay layer regardless of top and bottom drainage conditions.

注意，使用式(4-19)和式(4-20)计算沉降量时，H是黏土层的总厚度，不考虑顶部和底部的排水条件。

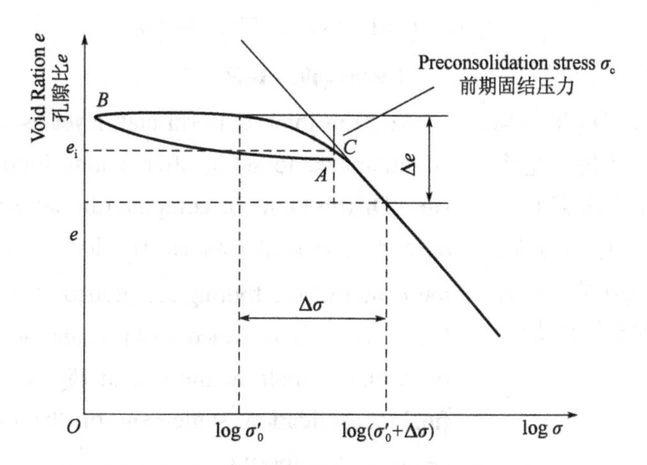

Figure 4-9 Settlement computation for overconsolidated soils.

图4-9 超固结土的沉降计算

4.3.2 多层土或厚土层的固结沉降

如图4-10所示,当土层较厚或有多种不同土层时,σ_0和$\sigma_0+\Delta\sigma$沿土层深度不是一个恒定值,此时采用式(4-19)和(4-20)进行沉降计算是不合理的。这种情况下,土层可被分为数层,根据第4.3.1节的方法,通过H_i、$\sigma_{0,i}$和$\Delta\sigma_i$值计算各层的最终沉降$s_{f,i}$,最终总沉降量s_f是$s_{f,i}$的总和,这种方法称为分层总和法。

4.3.2 Consolidation settlement for multilayers or a thick clay layer

When clay layers are thick or consist of several different clay layers, one-step computation by Equations (4-19) and (4-20) is not suitable, since σ_0 and $\Delta\sigma$ are not considered to be constant values throughout the depth of the clay layers, as seen in Figure 4-10. In this case, the whole clay layer is divided into several sublayers as seen in the figurc. Final settlement $s_{f,i}$ for each sublayer is computed from the methods described in Section 4.3.1, using H_i, $\sigma_{0,i}$ and $\Delta\sigma_i$ values, which can be obtained at the midpoints of each sublayer i. The final total settlement s_f is the summation of $s_{f,i}$. This method is called "layer wise summation method".

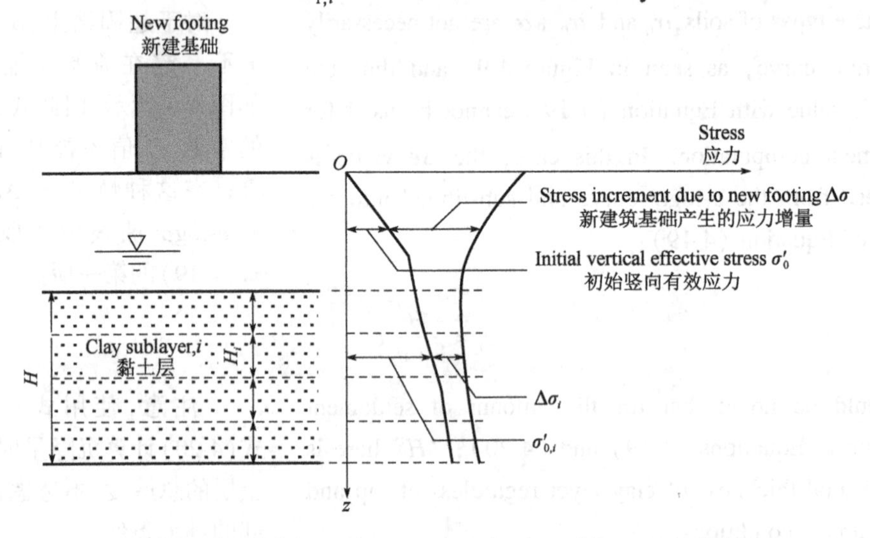

Figure 4-10 Consolidation settlement computation for multilayers or a thick layer

图4-10 多层土或厚层土的固结沉降计算

例题 4-2

图4-11为20 m厚的均匀黏土层,铺设基础后预计将产生沉降。图中绘制了初始垂直有效应力和新基础的附加应力分布,图4-12土体的e-logσ曲线,计算基础中心以下黏土层的最终固结沉降量。

Example 4-2

A 20 m thick uniform clay layer as shown in Figure 4-11 is anticipated to settle after a new footing is placed on the site. Distributions of computed initial vertical effective stress σ_0' and incremental stress $\Delta\sigma$ due to the new footing under the center of the footing are plotted also in the figure. The e-log σ curve is obtained from a laboratory consolidation test for a clay sample at the site in Figure 4-12. Compute total final consolidation settlement of the clay layer under the center of the footing.

解:

将200 m厚的黏土层分为厚

Solution

A 200 m thick clay layer is divided into four equal sublayers and σ'_0 and incremental stress $\Delta\sigma$ for each sublayer are read at the midpoints in Figure 4-11. Corresponding e_0 to σ'_0 and e_f to $\sigma'_0+\Delta\sigma$ of each sublayer are read from Figure 4-13 (enlarged version of Figure 4-12). The results are summarized in Table 4-1, where Equation (4-20) was used to compute $s_{f,i}$.

Thus, estimated total final settlement of the 20 m thick clay layer is 0.273 m.

Table 4-1 is based on utilization of the e-log σ curve to obtain Δe_i. If clays are normally consolidated, $s_{f.i}$ can be calculated from Equation (4-19).

度相等的四层，从图4-11可获取每层的σ'_0和附加应力$\Delta\sigma$，从图4-13(图4-12放大版)可获取每层土体对应的$e_0(\sigma'_0)$和$e_f(\sigma'_0+\Delta\sigma)$，通过公式(4-20)可计算$s_{f,i}$，结果如表4-1所示。

因此，20m厚黏土层的最终总沉降量为0.273m。

表4-1通过e-log σ曲线获得Δe_i。如果黏土是正常固结的，$s_{f.i}$可根据方程(4-19)计算。

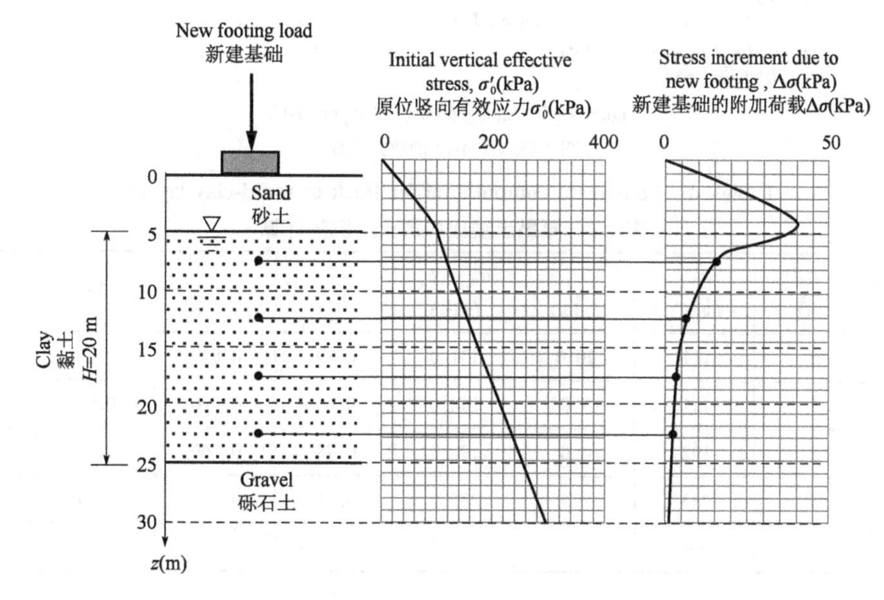

Figure 4-11 Diagram for Example 4-2

图4-11 例题4-2的附图

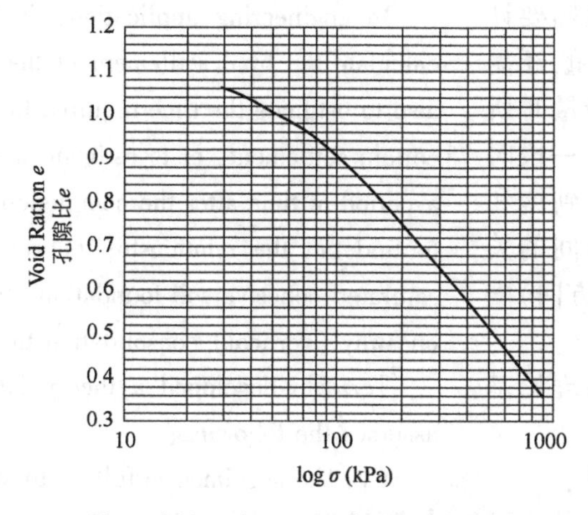

Figure 4-12 e-logσ curve for Example 4-2

图4-12 例题4-2 e-logσ的曲线

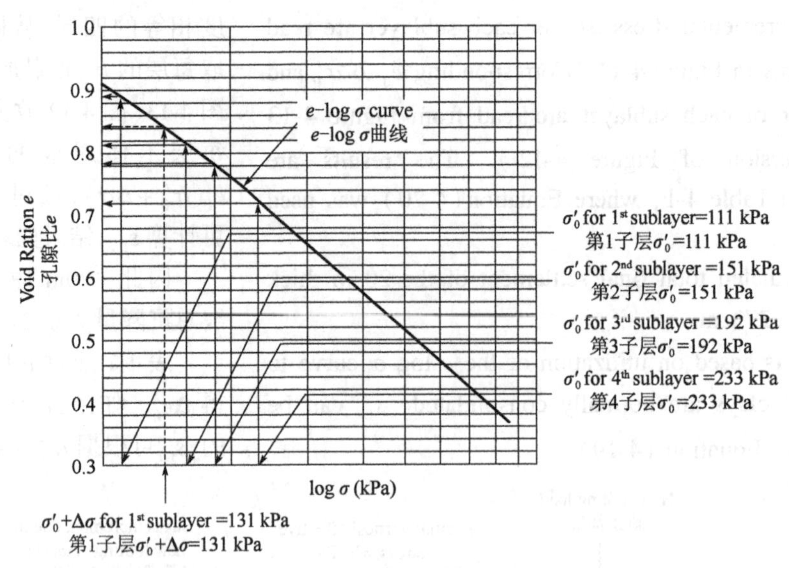

Figure 4-13 Enlarged curve of Figure 4-12

图 4-13 图 4-12 的放大曲线

Table 4-1 Settlement computation for thick or multi-clay layers

表 4-1 厚黏土层或多层土层沉降计算

i	H (m)	$\sigma_{0,i}$ (kPa)	σ_i (kPa)	$\sigma_{0,i}+\sigma_i$ (kPa)	$e_{0,i}$	$e_{f,i}$	Δe_i	$s_{f,i}$ (m)
1	5	111	20.0	131.0	0.89	0.840	0.05	0.132
2	5	151	7.0	158.0	0.81	0.790	0.02	0.055
3	5	192	4.0	196.0	0.78	0.760	0.02	0.056
4	5	233	2.5	235.5	0.72	0.714	0.01	0.030
Σ	20	—	—	—	—	—	—	0.273

4.4 太沙基一维固结理论

在工程实践中,不仅需要计算建筑基础的最终沉降量,还需计算基础达到某一沉降量时所需时间,或计算建筑完工一段时间后的沉降量。为了得到饱和黏性土地基沉降量与时间的关系,目前均以一维(竖直方向)固结理论为基础进行计算。

太沙基提出的固结理论基本假定包括:

(1)土是完全饱和的;

(2)土颗粒和孔隙水均不可压缩;

4.4 Terzaghi's consolidation theory

In engineering application, it is necessary not only to calculate the final settlement of the building foundation, but also to estimate the time required for the foundation to reach a certain settlement, or to estimate the possible settlement after a period of time after the completion of the building. In order to find out the relationship between settlement and time of saturated cohesive soil foundation, the calculation is based on one-way (vertical) consolidation theory.

Terzaghi developed a theory for consolidation model. It assumes the following:

(1) The specimen is fully saturated;

(2) Water and solid components are incompressible;

(3) Darcy's law is strictly applied;

(4) Flow of water is one dimensional.

Note that those assumptions are nearly all valid for one-dimensional consolidation for fully saturated soils. Figure 4-14 shows a three-phase diagram of a fully saturated soil. In the model, the original total volume is 1.0, and the original volume of void (full of water) is initial porosity, n_0.

(3)孔隙水渗流遵从达西定律;

(4)水的流动是单向的。

注意,这些假设适用于饱和土的一维固结。图4-14为饱和土的三相图,初始体积计为1.0,初始孔隙(充满水)体积为n_0。

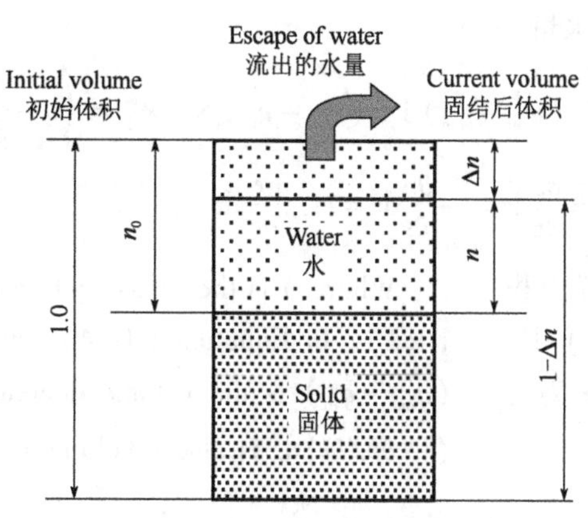

Figure 4-14 Three-phase model for consolidation process

图4-14 固结过程的三相示意图

During the consolidation process, when the effective stress increased from initial σ_0 to the current σ', water is squeezed out at the amount of Δn, and the current volume of void becomes n as seen. Thus:

在固结过程中,当有效应力从σ_0增大到σ'时,排出水的体积为Δn,此时孔隙体积变为n。因此:

$$\Delta n = n_0 - n = m_v \Delta\sigma' = m_v(\sigma' - \sigma'_0) \tag{4-21}$$

Where $\Delta\sigma'$ is the effective stress change, and m_v is defined as the coefficient of volume change, which is a parameter to connect the effective stress change to the volume change.

By taking the first derivative of Equation (4-21) with respect to time t:

式中,$\Delta\sigma'$为有效应力变化量;m_v为压缩系数,表示土体有效应力随体积变化的参数。

取式(4-21)对时间t的一阶导数:

$$\frac{\partial \Delta n}{\partial t} = \frac{\partial n_0}{\partial t} - \frac{\partial n}{\partial t} = 0 - \frac{\partial n}{\partial t} = m_v\left(\frac{\partial \sigma'}{\partial t} - \frac{\partial \sigma'_0}{\partial t}\right) = m_v\left(\frac{\partial \sigma'}{\partial t} - 0\right) \tag{4-22}$$

Then:

可得:

$$\frac{\partial n}{\partial t} = -m_v \frac{\partial \sigma'}{\partial t} \tag{4-23}$$

Figure 4-15 shows a square tube element with $1 \times 1 \times dz$ dimensions. Water flows toward the upper z direction, and the inflow water velocity v and the outflow velocity $v + (\partial v/\partial z)dz$ are shown. q_{in} and q_{out} are the inflow water rate and the outflow flow rate, respectively. By knowing that $q_{out} - q_{in}$ is the

图4-15为饱和黏土层任一深度处的微元体(断面积=1×1,厚度dz),水沿z坐标正方向流动,入口和出口流速分别为v和$v + (\partial v/\partial z)dz$。入口和出口

在单位时间内流过的水量分别为 q_{in} 和 q_{out}。显然 $q_{out}-q_{in}$ 是 $1\times1\times dz$ 水流总体积在单位时间内的变化,式(4-22)中 $\partial n/\partial t$ 也是水流总体积在单位时间内的变化。即在 dt 时间内,微元体中孔隙体积的变化等于同一时间内从微元体中流出的水量。

volume change per unit time for $1\times1\times dz$ total volume, and that $\partial n/\partial t$ in Equation (4-22) is also the volume change per unit time for total volume. That is, within dt, the change in pore volume in the microelement is equal to the amount of water flowing out of the microelement at the same time.

$$q_{out}-q_{in}=(v_{out}-v_{in})A=(v_{out}-v_{in})\times1\times1=\left(v+\frac{\partial v}{\partial z}dz\right)-v$$
$$=\frac{\partial v}{\partial z}dz=-\frac{\partial n}{\partial t}dz=m_v\frac{\partial \sigma'}{\partial t}dz \tag{4-24}$$

式中,A 为水流的横截面积(1×1)。在公式(4-24)中,($q_{out}-q_{in}$)和($-\partial n/\partial t$)dz 为正值时代表土体积减小。

Where A is the cross-sectional area for water flow (i. e., 1×1). In Equation (4-24), note that the positive value of ($q_{out}-q_{in}$) is the volume decrease and the positive value of ($-\partial n/\partial t$)dz is also a volume decrease.

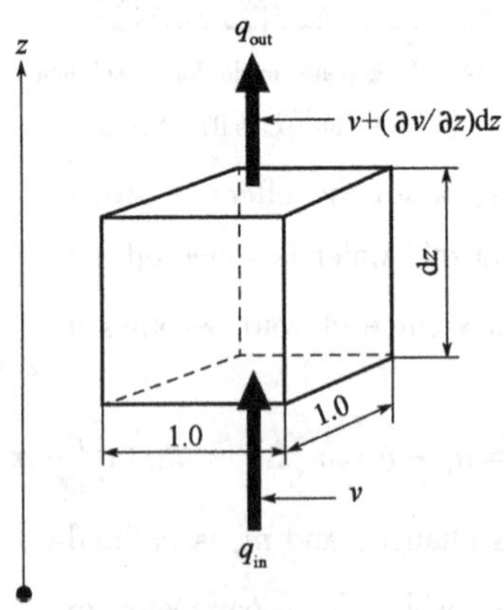

Figure 4-15 Vertical water flow through a square tube ($1\times1\times dz$)

图 4-15 流经方形管的垂直水流($1\times1\times dz$)

由式(4-24)可得式(4-25):

From Equation (4-24), Equation (4-25) is obtained:

$$\frac{\partial v}{\partial z}=m_v\frac{\partial \sigma'}{\partial t} \tag{4-25}$$

有效应力为 $\sigma'=\sigma-u$,对其求时间 t 的一阶导数,可得:

Now, the effective stress is defined as $\sigma'=\sigma-u$, and taking the first derivative with respect to time t, then:

$$\frac{\partial \sigma'}{\partial t}=\frac{\partial \sigma}{\partial t}-\frac{\partial u}{\partial t}=0-\frac{\partial u}{\partial t}=-\frac{\partial u}{\partial t} \tag{4-26}$$

式中,固结过程中施加的总应力 σ 是恒定的,因此 $\partial\sigma/\partial t=0$,可得:

Where $\partial\sigma/\partial t=0$ since the applied total stress σ is constant during the consolidation process. Thus:

$$\frac{\partial \sigma'}{\partial t}=-\frac{\partial u}{\partial t} \tag{4-27}$$

Here, Darcy's law for water flow is introduced:

根据达西定律：

$$v = ki = k\frac{-\partial h_{\mathrm{p}}}{\partial z} = k\frac{-\partial\left(\frac{u}{\gamma_{\mathrm{w}}}\right)}{\partial z} = -\frac{k}{\gamma_{\mathrm{w}}}\frac{\partial u}{\partial z} \tag{4-28}$$

Where k is the coefficient of permeability, and i is the hydraulic gradient. ∂h_{p} is the pressure head difference and is negative for positive water flow velocity v in Figure 4-15.

式中，k 为渗透系数；i 为水力梯度；∂h_{p}为水头差，当图4-15 中水流速度 v 为正时，∂h_{p}为负值。

From Equation (4-28):

根据式(4-28)可得：

$$\frac{\partial v}{\partial z} = -\frac{k}{\gamma_w}\frac{\partial^2 u}{\partial z^2} \tag{4-29}$$

By equating Equations (4-25) and (4-29) and by substituting Equation (4-27):

将式(4-25)与式(4-29)代入式(4-27)可得：

$$\frac{\partial u}{\partial t} = \frac{k}{m_{\mathrm{v}}\gamma_{\mathrm{w}}}\frac{\partial^2 u}{\partial z^2} = C_{\mathrm{v}}\frac{\partial^2 u}{\partial z^2} \tag{4-30}$$

$$C_{\mathrm{v}} = \frac{k}{m_{\mathrm{v}}\gamma_{\mathrm{w}}} \tag{4-31}$$

Equation (4-30) is called the consolidation equation. C_{v} is the coefficient of consolidation with a unit of length^2/time ($\mathrm{m^2/s}$ etc.) and is a key material parameter in consolidation theory.

式(4-30)为饱和黏性土的渗流固结微分方程。C_{v} 为固结系数，单位 $\mathrm{m^2/s}$，是太沙基固结理论的一个关键物理参数。

To solve the second order of partial differential Equation (4-30), four boundary (and initial) conditions are required. Figure 4-16a) plots the pore water pressure u with depth z as a faction of time t. The top and bottom layers are assigned as drainage layers like sand or gravel, and a clay layer ($2H$ thickness) is sandwiched between them. Excess pore water pressure can only be drained through the drainage layers, and thus at the mid-depth H, the highest pore water pressure remains for $0 < t < \infty$ as seen. The initial and boundary conditions for this case are:

求解二阶偏微分方程(4-30)需要四个边界条件（和初始条件）。图4-16a)表示时间 t 时深度 z 处土体的孔隙水压力为 u。顶层和底层为排水层，如砂土或碎石土，中间夹一层厚度为 $2H$ 的黏土。超静孔隙水压力只能通过排水层消散，因此在中间深度 H 处，任何时刻均为最大孔隙水压力。此时的初始条件和边界条件包括：

(1) u(at any z, at $t=0$) $=\Delta\sigma$;

(2) u(at any z, at $t=\infty$) $=0$;

(3) u(at $z=0$, at any t) $=0$;

(4) u(at $z=2H$, at any t) $=0$.

(1) $u(z, t=0) = \Delta\sigma$；

(2) $u(z, t=\infty) = 0$；

(3) $u(z=0, t) = 0$；

(4) $u(z=2H, t) = 0$。

These conditions can also be applied to Figure 4-16b), where the bottom layer is impervious, so that water drainage occurs only at the top boundary. In this case, the clay thickness

上述条件也适用于图4-16b)，即底层不透水、仅顶部边界排水的情况。在这种情况下，黏土厚

度视为 H,此时底部边界条件为 $\partial u/\partial z=0$。

is treated as H and the bottom boundary condition is $\partial u/\partial z=0$.

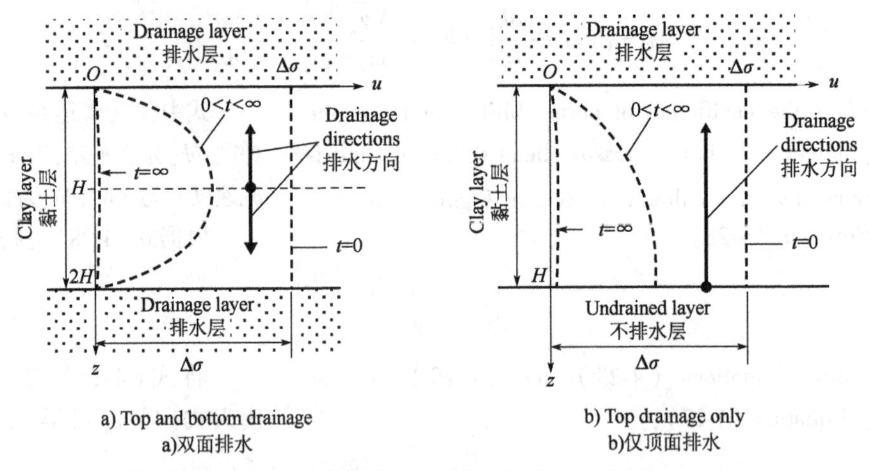

Figure 4-16 Initial and boundary conditions for the consolidation equation

图 4-16 固结方程的初始和边界条件

根据初始条件和边界条件,可得式(4-30)的解为:

By using the initial and boundary conditions, the following solution of Equation (4-30) is obtained:

$$u(z,t)=\frac{4}{\pi}\Delta\sigma\sum_{n=0}^{\infty}\left[\frac{1}{2n+1}\sin\frac{(2n+1)\pi z}{2H}\cdot e^{-\frac{(2n+1)^2\pi^2C_v}{4H^2}t}\right] \tag{4-32}$$

代入 $n=0$ 及几个更大的数值,得到给定 z 和 t 的数值解。

By substituting $n=0$ to several higher values, the solution converges and the numerical solution is obtained for given z and t.

为了使计算更加简便,引入时间因数 T_v:

To make an operation much simpler, time factor T_v is introduced as:

$$T_v=\frac{C_v t}{H^2} \tag{4-33}$$

T_v是一个无量纲变量,表示与土固结系数 C_v 和排水距离 H 相关的时间因子。H 应取到排水层的最大距离,如图 4-16 所示。

This is a nondimensional variable to express the time relative to material parameter C_v and drainage distance H. In this equation, H should be taken as the longest distance to the drainage layer, as seen in Figure 4-16.

将时间因子 T_v 表达式(4-33)代入式(4-32),可得:

When T_v is substituted into Equation (4-33), it becomes:

$$\begin{aligned}u(z,t)&=\frac{4}{\pi}\Delta\sigma\sum_{n=0}^{\infty}\left[\frac{1}{2n+1}\sin\frac{(2n+1)\pi z}{2H}\cdot e^{-\frac{(2n+1)^2\pi^2}{4}T_v}\right]\\&=f\left(\Delta\sigma,\frac{z}{2H},T_v\right)\end{aligned} \tag{4-34}$$

在式(4-34)中,孔隙水压力 u 表示为三个独立参数的函数:$\Delta\sigma$、$z/2H$ 和 T。

In this equation form, the pore water pressure u is expressed as a function of three independent parameters: $\Delta\sigma$, $z/2H$, and T.

By referring to the three-phase diagram of Figure 4-9 and using Equation (4-21), the final consolidation settlement s_f (at $t=\infty$) for a clay layer of thickness H can obtained as:

参考图 4-14 的三相图及方程(4-21),厚度为 H 的黏土层的最终固结沉降量 s_f($t=\infty$ 时)为:

$$s_f=\Delta n_f \cdot H=m_V\Delta\sigma'_f H=m_v\Delta\sigma H \tag{4-35}$$

In the preceding expression, subscript "f" stands for "final". Meanwhile, the settlement s_t at any arbitrary time, t, is obtained from an integration of settlement $\Delta n\times \mathrm{d}z$ for a small clay thickness dz over the total clay layer thickness H as seen in Figure 4-17.

式中,下标"f"代表"final"。同时,任意时间 t 的沉降量 s_t 通过厚度为 dz 微元体的沉降 $\Delta n\times$ dz 与黏土层厚度 H 积分得到,如图 4-17 所示。

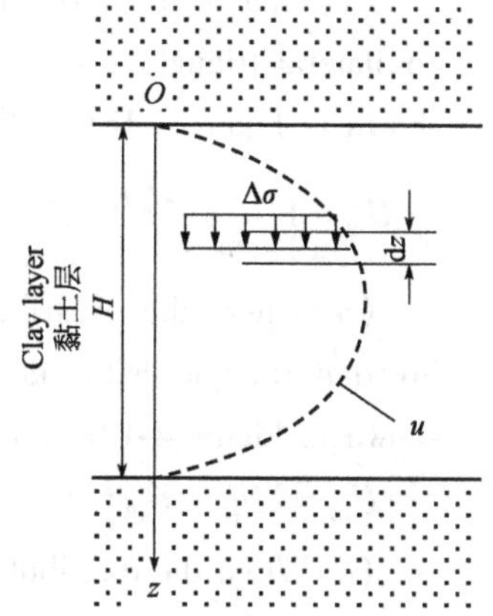

Figure 4-17 Settlement computation model

图 4-17 沉降计算模型

Thus:

因此:

$$s_t=\int_0^H\Delta n\mathrm{d}z=\int_0^H m_v\Delta\sigma'\mathrm{d}z=m_v\int_0^H(\Delta\sigma-u)\mathrm{d}z=m_v\Delta\sigma H-m_v\int_0^H u\mathrm{d}z \tag{4-36}$$

Where $\Delta\sigma$ is the increased stress at depth z. Since u is given in Equation (4-34), Equation (4-36) becomes:

式中,$\Delta\sigma$ 为深度 z 处的附加应力。孔隙水压力 u 在式(4-34)中计算得到,那么式(4-36)变为:

$$s_t=m_v\Delta\sigma H\left[1-\frac{8}{\pi^2}\sum_{n=0}^{\infty}\left(\frac{1}{2n+1}\cdot e^{-\frac{(2n+1)^2\pi^2}{4}T_v}\right)\right] \tag{4-37}$$

Now, the degree of consolidation U is defined as the percentage of settlement at an arbitrary time t to its final settlement at $t=\infty$ and it is computed from Equations (4-35) and (4-37) as:

固结度 U 为地基在任意时间 t 的沉降量与最终沉降量之比,可由式(4-35)和(4-37)计算得到:

$$U_t=\frac{s_t}{s_f}=\left[1-\frac{8}{\pi^2}\sum_{n=0}^{\infty}\left(\frac{1}{2n+1}\cdot e^{-\frac{(2n+1)^2\pi^2}{4}T_v}\right)\right]=f(T_v) \tag{4-38}$$

As seen in Equation (4-38), the degree of consolidation U_t is only a function of time factor T_v. The corresponding $U_t=f(t)$ relation in various cases can be obtained.

根据式(4-38)可知,固结度 U_t 是时间因子 T_v 的函数。可得到不同情况下相应的 $U_t=f(t)$ 的关系式。

(1)附加应力图形为:①矩形,单面或双面排水;②三角形、梯形,双面排水情况,如图4-18a)所示,它们的固结度计算方法均为式(4-38)。由于括号内的级数收敛很快,使用时可取第一项,即:

(1) When the distribution of additional stress is ①rectangular with single or double drainage surface; ②the distribution of additional stress is triangle, trapezoid with double drainage surfaces, as shown in Figure 4-18a), the degree of consolidation all can be obtained through Equation (4-38). Since the series in brackets converge quickly, the first term can be used in practice, that is:

$$U_{tA} = 1 - \frac{8}{\pi^2} e^{-\frac{\pi^2}{4} T_v} \tag{4-39}$$

(2)附加应力图形为正三角形,单面排水(排水面在上),如图4-18b)所示,其固结度为:

(2) When the distribution of additional stress is equilateral triangle with only an upper drainage surface, as shown in Figure 4-18b), the degree of consolidation is:

$$U_{tB} = 1 - \frac{32}{\pi^2} e^{-\frac{\pi^2}{4} T_v} \tag{4-40}$$

(3)附加应力图形为倒三角形,单面排水(排水面在上),如图4-18c)所示,其固结度为:

(3) When the distribution of additional stress is an inverted triangle with only an upper drainage surface, as shown in Figure 4-18c), the degree of consolidation is:

$$U_{tC} = 2U_{tA} - U_{tB} \tag{4-41}$$

(4)附加应力图形为梯形,单面排水,如图4-18d)所示,其固结度为:

(4) When the distribution of additional stress is trapezoid with single drainage surface, as shown in Fig 4-18d), and the degree of consolidation is:

$$U_{tD} = U_{tA} + \frac{a-1}{a+1}(U_{tA} - U_{tB}) \tag{4-42}$$

式中,$a = \sigma_a / \sigma_b$;σ_a为排水面的附加应力;σ_b为不排水面的附加应力。

Where $a = \sigma_a / \sigma_b$, σ_a is the additional stress on the drainage surface, and σ_b is the additional stress on the undrainage surface.

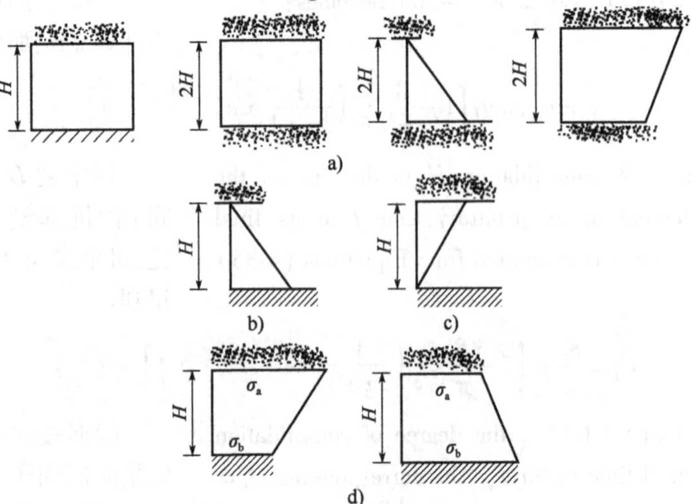

Figure 4-18 Different consolidation conditions

图4-18 不同的固结条件

The relationship between the degree of consolidation U_t and their time factor T_v in the above four cases can be referred to Table 4-2.

上述四种情况的固结度 U_t 和它们时间因数 T_v 的关系，可参考表4-2。

Table 4-2 $U_t = U_t(T_v)$ in different cases

表4-2 不同情况下的 $U_t = U_t(T_v)$

T_v	Degree of consolidation 固结度				T_v	Degree of consolidation 固结度			
	U_{tA}	U_{tB}	U_{tC}	$U_{tA} \cdot U_{tB}$		U_{tA}	U_{tB}	U_{tC}	$U_{tA} \cdot U_{tB}$
0.004	0.080	0.008	0.152	0.072	0.20	0.504	0.370	0.638	0.134
0.008	0.104	0.016	0.192	0.088	0.25	0.562	0.443	0.682	0.119
0.012	0.125	0.024	0.226	0.101	0.30	0.613	0.508	0.719	0.105
0.020	0.160	0.040	0.286	0.120	0.35	0.658	0.565	0.752	0.093
0.028	0.189	0.056	0.322	0.133	0.40	0.698	0.615	0.780	0.083
0.036	0.214	0.072	0.352	0.142	0.50	0.764	0.700	0.829	0.064
0.048	0.247	0.095	0.398	0.152	0.60	0.816	0.765	0.866	0.051
0.060	0.276	0.120	0.433	0.156	0.80	0.887	0.857	0.918	0.030
0.072	0.303	0.144	0.462	0.159	1.00	0.931	0.913	0.949	0.018
0.100	0.357	0.198	0.516	0.159	2.00	0.994	0.993	0.995	0.001
0.125	0.399	0.244	0.554	0.155	∞	1.000	1.000	1.000	0.000
0.167	0.461	0.318	0.605	0.143					

Example 4-3

In a laboratory consolidation test, a 12.7 mm thick clay specimen was tested with top and bottom drained condition, and 90% consolidation was accomplished in 15.8 min (t_{90} = 15.8 min). In the field, the same clay material with the thickness of 6.5 m is sandwiched by top sand and bottom gravel layers for drainage. How long does the field clay take to accomplish 50% and 90% consolidation, respectively?

例题 4-3

对厚度12.7 mm的黏土试样进行顶部和底部排水的室内固结试验，在15.8 min（t_{90} = 15.8 min）内完成了90%的固结。现场厚度为6.5 m的相同黏土层，其顶部为砂土层，底部为砾石层，分别需要多长时间才能完成50%和90%的固结？

Solution

In the laboratory test, top and bottom are drainage layers, so the clay thickness 12.7 mm = $2H$ and T_{90} = 0.848 from Table 4-1. Inserting these values into Equation (4-33):

解：

室内试验中，顶部和底部均为排水层，黏土厚度为12.7 mm = $2H$，由表4-1可知 T_{90} = 0.848。将这些值代入式(4-33)，有：

$$C_v = \frac{H^2}{t_{90}} T_{90} = \frac{\left(\frac{12.7}{2}\right)^2}{15.8} \times 0.848 = 2.164(\text{mm}^2/\text{min})$$

From the field drainage condition, $2H$ = 6.5 m. Also T_{50} = 0.197 from Table 4-1. Utilizing Equation (4-33), 50% consolidation time, t_{50}, is:

根据现场排水条件，$2H$ = 6.5 m。表4-1中的 T_{50} = 0.197。利用式(4-33)，达到50%固结度所需的时间 t_{50} 为：

$$t_{50}=\frac{H^2}{C_v}T_{50}=\frac{\left(\frac{6.5\times1000}{2}\right)^2}{2.164}\times0.197=9.615\times10^5=667.7(\mathrm{d})$$

同样,对于达到90%固结度所需的时间 t_{90} 为:

Similarly, for 90% consolidation time t_{90}:

$$t_{90}=\frac{H^2}{C_v}T_{90}=\frac{\left(\frac{6.5\times1000}{2}\right)^2}{2.164}0.848=41.39\times10^5(\mathrm{min})=2874(\mathrm{d})=7.87(\mathrm{a})$$

或者,根据式(4-33)得 C_v 为:

Or, from Equation (4-33) and by using a common C_v value:

$$C_v=\frac{H^2}{t_{50}}T_{50}=\frac{H^2}{t_{90}}T_{90}$$

$$t_{90}=\frac{T_{90}}{T_{50}}t_{50}=\frac{0.848}{0.197}\times667.7=4.305\times667.7=2874(\mathrm{d})=7.87(\mathrm{a})$$

习　题

(1)设土样厚3 cm,在100~200 kPa范围内的压缩系数 $a=2\times10^{-4}$,当压力为100 kPa时,$e=0.7$。求:①土样的无侧向膨胀变形模量;②土样压力由100 kPa加到200 kPa时,土样的压缩量 S。

(2)某饱和土层厚3 m,上下两面透水,在其中部取土样进行固结试验(试样厚2 cm),在20 min后固结度达50%,求:

①固结系数 C_v;

②该土层在满布压力作用下,达到90%固结度所需的时间。

(3)设有一宽3 m的条形基础,基底以下地层依次为2 m砂层、3 m饱和软黏土层、不透水的岩层。试求:

①取原状饱和黏土样进行固结试验,试样厚2 m,试样上表面排水,测得固结度为90%时所需时间为5 h,求其固结系数;

Excercises

(1) Suppose that the thickness of soil sample is 3 cm, the coefficient of compressibility in the pressure range of 100~200 kPa is 2×10^{-4}, and $e=0.7$ when the pressure is 100 kPa. Calculate: ① deformation modulus without lateral expansion of soil sample; ② compression S of soil sample when the pressure of soil sample increases from 100 kPa to 200 kPa.

(2) A saturated soil layer is 3 m thick and permeable on both sides. A soil sample is taken from the middle part of the saturated soil layer, and the consolidation test is carried out in thc laboratory (the sample is 2 cm thick). After 20 minutes, the degree of consolidation reaches 50%. Calculate:

① Consolidation coefficient C_v;

② The time required for the soil layer to reach consolidation degree of 90% under the action of full pressure.

(3) There is a strip foundation with a width of 3 m, under which there is a 2 m sand layer, a 3 m thick saturated soft clay layer, and an impermeable rock layer. Calculate:

① Take the undisturbed saturated clay sample for consolidation test, the thickness of the sample is 2 m, upper drainage of the sample, the required time for a measured consolidation degree of 90% is 5 h, calculate the consolidation coefficient;

② The foundation load is added at one time. After how much time, the saturated clay layer will complete 60% of the total settlement.

②基础荷载是一次加上的，问经过多少时间饱和黏土层完成总沉降量的60%。

Chapter 5 Shear Strength of Soils 土的抗剪强度

5.1 土的直剪试验与库仑定律

5.1.1 直剪试验

土体强度的两种来源包括：摩擦阻力和沿剪切带的黏聚阻力。图5-1a)中土体单元受法向应力和剪切应力的作用土体的剪切过程,可将其视为如5-1b)所示的表面粗糙的块体和板体模型。在该模型中,剪应力τ由块体之间的摩擦力和黏聚力c来抵抗。

5.1 Direct shear test and Coulomb's Law

5.1.1 Direct shear test

Soil strength may be attributed to two different mechanisms of the material: its frictional resistance and the cohesive resistance along the shearing zone. The shearing of a soil assemblage that is subjected to normal stress and shear stress [Figure 5-1a)] can be modeled as a block on a solid plate with a rough surface, as shown in Figure 5-1b). In the model, the shear stress τ is resisted by a frictional mechanism and by cohesion c between the interface of the block and the solid plate.

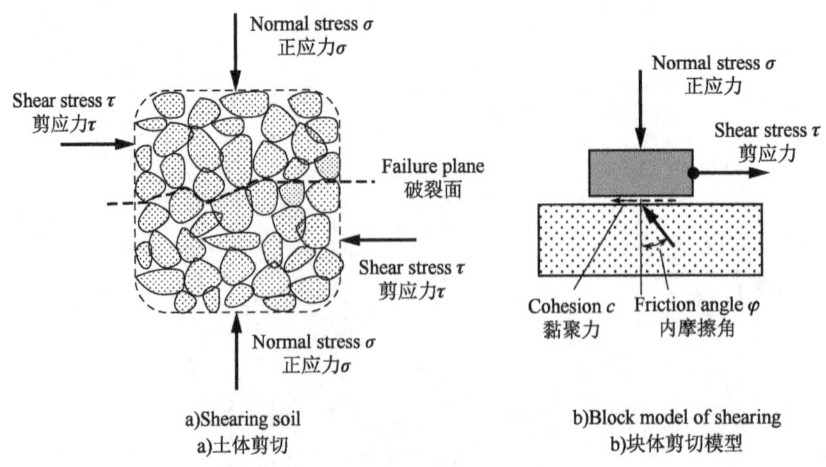

Figure 5-1 Shearing in soil mass

图5-1 土的剪切

图5-2是测量土体抗剪强度参数的装置,它由上、下两个剪

Figure 5-2 shows a model of the basic type of device used to determine soil strength parameters. It consists of upper and lower shear boxes, with a soil specimen placed inside the box. A vertical normal force F_v and hence a normal stress σ ($=F_v$/specimen area) is applied and held constant. In most such devices, the upper box is fixed, and the lower box is movable on low-friction rollers at the base. The lower box is pulled or pushed to apply a horizontal thrust (shear force) T, and thus, a shear stress τ ($=T$/specimen area) is induced along the middle plane of the specimen.

切盒组成,土样放在剪切盒内。当施加竖向压力F_v时,竖向应力σ($=F_v$/试样面积)保持恒定。通常,上剪切盒位置固定,对下剪切盒施加水平推力(剪切力)T,在底座的低摩擦滚轴上移动。试样中部剪切面上的剪应力为$\tau=T$/试样面积。

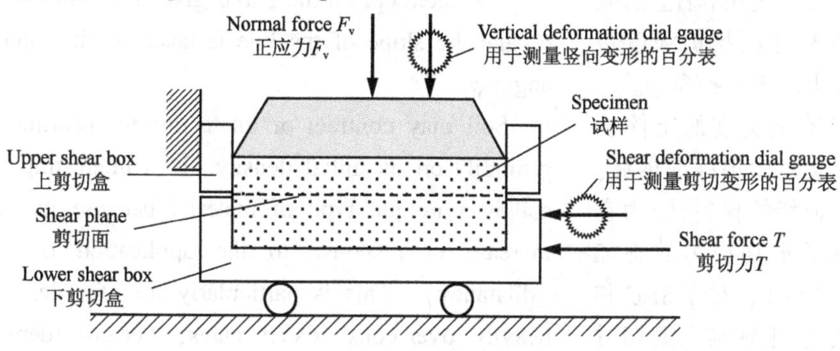

Figure 5-2 Direct shear test setup

图 5-2 直剪仪

For a given soil with a similar density, several direct shear tests are conducted under different normal stresses. Peak shear strength values τ_f are measured for each test. Then σ and τ_f relations are plotted as in Figure 5-3. A linear relation is obtained through the data points.

在不同竖向压力下对同一种土进行直剪试验,得出不同σ对应的抗剪强度τ_f,并绘制关系图。如图 5-3 所示,其结果为一条直线,该直线即为抗剪强度包线。

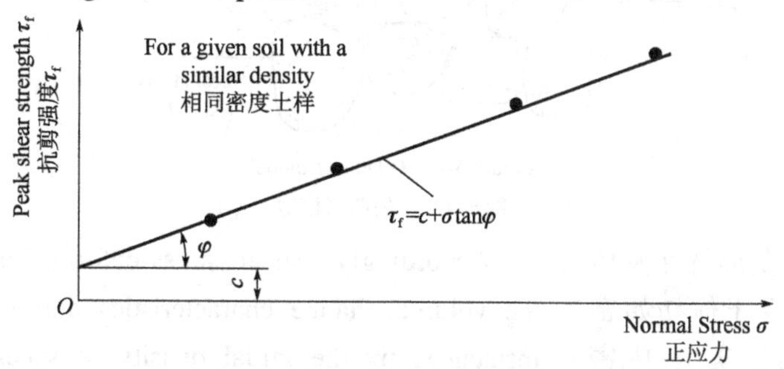

Figure 5-3 Determination of c and φ from direct shear tests

图 5-3 直剪仪确定 c、φ 值

5.1.2 Coulomb's Law

The angle φ can be interpreted as the friction angle between facing soil elements along the shear surface. The cohesion c is called the cohesion of the soil. In the block model (Figure 5-1), the cohesion resistance can be simulated using a coating of heavy grease between the block and the plate, and thus, it is independent of the applied

5.1.2 库仑定律

剪切面上土体颗粒之间的摩擦作用可用摩擦角φ表示,土颗粒之间的黏聚作用称为土的黏聚力c。如图 5-1 所示,块体和板体之间的黏聚作用可由油脂土层进行模拟,其大小与法向

应力σ无关。黏聚力主要是土粒间的相互作用力,是细粒土(黏土等)的特有属性。

normal stress σ. In soils, normal stress-independent cohesion results from particle-to-particle close-range interactive forces and is a material property of fine-grained soils (clays or other cohesive soils).

相应地,土体剪切破坏时的剪切强度为:

Accordingly, the total shear stress at failure τ_f is expressed as:

$$\tau_f = c + \sigma\tan\varphi \tag{5-1}$$

抗剪强度包线在τ轴上的截距为土的黏聚力c,抗剪强度包线的斜率为土的内摩擦角φ。

The intercept on the τ axis gives the cohesion component c, and the slope of the line is taken as the internal friction angle φ.

不同初始压实度的土体在剪切过程中会发生剪缩或剪胀。土体是一种独特的材料,因为它的体积会随着剪切应力的施加而增大(膨胀性)。对于密砂和超固结黏性土,土颗粒在剪切过程中会发生平动或滚动,土颗粒相对位置会发生变化,由此导致体积增大(剪胀性),如图5-4所示。

Soil may contract or dilate during shearing, depending primarily on its initial density. It is interesting to note that soil is a unique type of material because its volume may increase in response to the application of shear stress (dilatancy). This is particularly true for dense sands and heavily over-consolidated clays, because densely packed grains or particles have to move or roll over neighboring grains to change their relative positions during shearing, as illustrated in Figure 5-4.

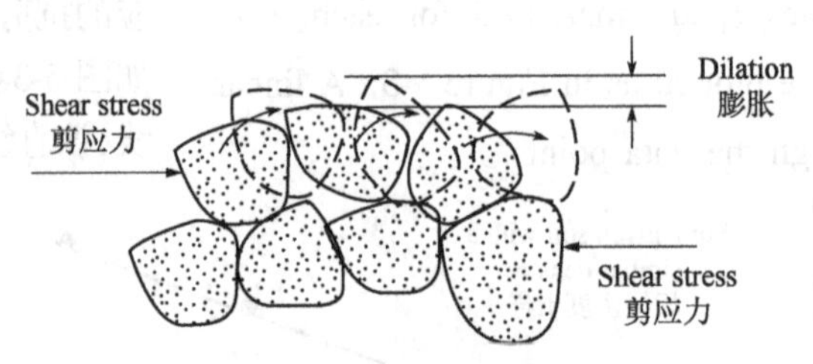

Figure 5-4 Dilatancy model

图5-4 土的剪胀模型

土的剪应力-变形特性及体积变化特性主要受土体初始密度影响,图5-5为密实土、中密土和松散土的剪应力-变形曲线,在大剪切变形时,不同密实度的土体剪应力-变形曲线均收敛于残余剪切强度,同时孔隙比也趋于一个稳定值。

Accordingly, shear stress-deformation relationship and the volume change characteristics during shear are largely influenced by the initial density of specimens. Figure 5-5 shows these relations for dense, medium-dense, and loose soils. As seen in the figure, the shear stress-deformation curves converge to the residual shear strength at a large shear deformation level. The void ratios also converge to a certain value at a large shear deformation level.

大变形阶段是指土体某部分(剪切面)的剪切变形非常大,该部分土颗粒从初始松散或密实状态向新的稳定状态转变,土

When a soil assemblage is sheared at a large deformation level, certain zones within the specimen (shear plane) are subjected to large shear deformation. Along these shear zones, where shear failure occurs, particles are oriented into a preferred direction, which changes from their original loose or dense configurations, and a steady-state flow (failure) mechanism is created. Therefore, at large shear deformation levels, all strengths converge to the residual strength, and all void ratios converge to a certain value, regardless of the original density of the soil assemblage. As illustrated in Figure 5-5b), initially dense soils undergo initial contraction and then dilation. In contrast, loose soils contract until failure. There exists an intermediate specimen density at which the void ratio remains nearly the same during shear. This void ratio is called the critical void ratio. A specimen at this void ratio does not contract or dilate during shear.

体发生剪切破坏。此时,不同初始密实度的土体的剪切强度均趋向于残余强度,孔隙比均趋向于稳定值。如图 5-5b)所示,剪切过程中,密实土体首先发生剪缩现象,然后发生剪胀现象;松散土体仅发生剪缩现象。在剪切过程中,存在一种密实度状态,此时土体孔隙比保持不变,该孔隙比称为临界孔隙比,此时土体既不剪缩也不剪胀。

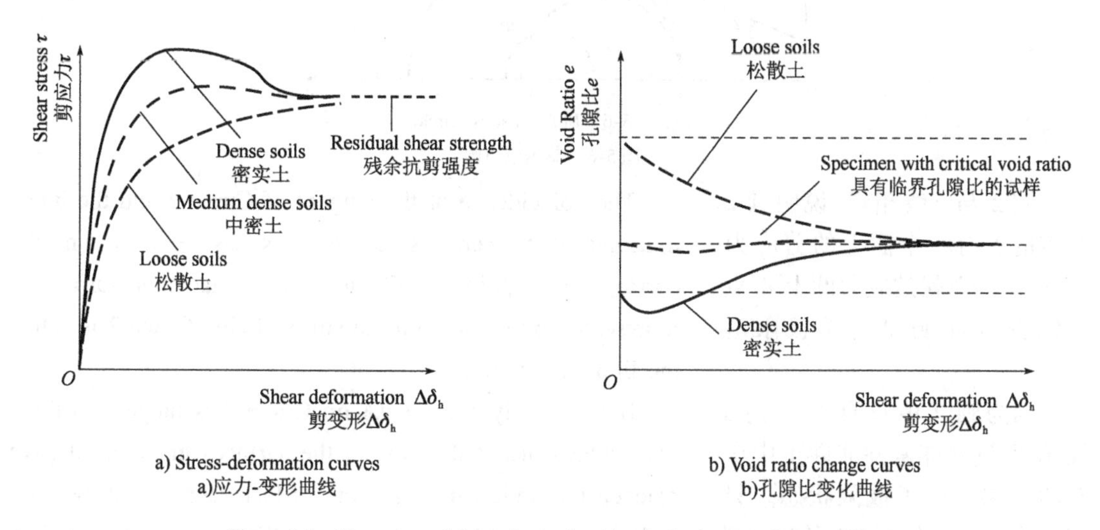

Figure 5-5 Shear stress-deformation and void ratio for loose and dense soils

图 5-5 松散土和密实土的剪切变形特性

5.2 Mohr's stress state

When the shear stress τ of a point in the soil is equal to the shear strength s, the soil is considered to be at the limit equilibrium state, and the stress state at this point is called the limit equilibrium stress state. Certain conditions, such as $\tau = s$, are applicable at the limit equilibrium stress state, and the limit equilibrium stress state can be expressed by the limit stress circle.

If the stress circle of a point and the shear strength line of a soil are drawn on the same σ-τ coordinate plane, we can judge whether a stress state is at the limit stress state.

5.2 莫尔应力状态

当土中一点的剪应力τ等于抗剪强度 s 时,土体被认为达到极限平衡状态,该点的应力状态称为极限平衡应力状态。极限平衡应力状态必须具备一定的条件,如上述的 $\tau = s$ 就是其中之一,除此之外还可以用极限应力圆来表示。

如果将土体一点的应力圆和土的抗剪强度包线绘制在同一 σ-τ坐标上,就可以判断该点的应力状态是否处于极限应力状态。如图5-6所示,有三个应力圆,其中圆1位于抗剪强度包线以下,说明圆1对应点的任何一个平面上的剪应力τ均小于其强度值s,因此土中这一点的应力状态未达到极限应力状态,土处于弹性平衡状态。

Figure 5-6 shows three stress circles, of which Circle 1 is located below the shear strength line, indicating that the shear stress τ on any plane of the point corresponding to Circle 1 is less than its strength value s. Therefore, the stress state at this point in the sand is not at the ultimate stress state but rather is in an elastic equilibrium state.

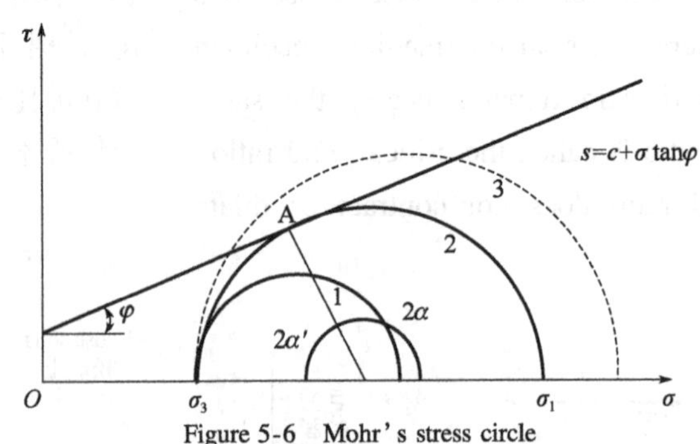

Figure 5-6 Mohr's stress circle

图5-6 莫尔应力圆

圆2与s线相切,说明圆2对应的点有一个面上的剪应力$\tau=s$,这个面就是应力圆上A点所代表的面,圆2称为极限应力圆。

The coincidence of the tangent of Circle 2 and the line s indicates that there is a shear stress $\tau = s$ on the corresponding point of Circle 2 and that this surface is represented by point A on the stress circle. Circle 2 is called the limit stress circle.

因为圆2中只有点A与s线相切,这意味着并非砂土中所有的面都达到了极限状态。只有A点所代表的面,其应力达到了极限状态,因此A点代表的面称为破裂面,破裂面与应力圆上的大小主应力面(σ_1和σ_3作用面)在应力圆上的夹角为2α和$2\alpha'$,实际上夹角为α和α',从图中可见:$2\alpha=90°+\varphi$,$2\alpha'=90°-\varphi$。因此:

Because only one pointA on Circle 2 is tangent to the s line, which means that not all the surfaces in the sand have reached the limit state, only the surface represented by point a has reached the limit state. Therefore, the surface represented by point a is called the failure surface. The angles between the failure surface and the major and minor principal stress surfaces (the surfaces acted on by σ_1 and σ_3) on the stress circle are 2α and $2\alpha'$, and the included angles are α and α'. It can be seen from Figure 5-6 that $2\alpha = 90° + \varphi$ and $2\alpha' = 90° - \varphi$. Hence:

$$\alpha = 45° + \frac{\varphi}{2} \tag{5-2}$$

$$\alpha' = 45° - \frac{\varphi}{2} \tag{5-3}$$

圆3与抗剪强度包线s相割,说明圆3所表示的砂土中的点已达到极限状态,并产生了塑

Circle 3 intersects the shear strength line s, which indicates that the point in the sand represented by Circle 3 has already reached the limit state and that plastic deformation has occurred. In fact, the soil has failed; therefore, Circle 3 is fictitious. Circle 3 is drawn only to compare the stress state of the soil; Circle 3 does not exist.

性变形。实际上,土早已发生了破坏,圆3只是为了比较土的应力状态而虚构的。

5.3 Mohr-Coulomb failure criterion

5.3 莫尔-库仑破坏准则

Equation (5-1), which is called the Mohr-Coulomb failure criterion, represents Coulomb's contribution to frictional law and the linear representation of its relation as well as Mohr's contribution to defining failure by a unique combination of normal stress σ and shear stress τ.

方程(5-1)称为莫尔-库仑破坏准则,代表库仑对摩擦定律的贡献和其线性关系表达式以及莫尔对土体破坏的定义及其所提出的正应力σ与剪应力τ的应力状态表达式。

Mohr-Coulomb shear strength theory can be summarized as:

莫尔-库仑抗剪强度理论总结如下:

(1) The shear strength on a plane is a function of the normal stress on the plane;

(1)破裂面上,材料的抗剪强度是法向应力的函数;

(2) The function $\tau_f = f(\sigma)$ can be expressed by a straight line over a certain stress range;

(2)在一定的应力水平范围内,$\tau_f = f(\sigma)$可近似简化为线性函数;

(3) If the shear stress at a point on one plane has reached the shear strength of the soil, failure has occurred at that point.

(3)当土体中任何一个面上的剪应力等于抗剪强度时,该点便发生破坏。

Terzaghi modified the Mohr-Coulomb equation to include his effective stress concept as:

太沙基采用其提出的有效应力概念修正了莫尔-库仑抗剪强度理论:

$$\tau_f = c' + \sigma'\tan\varphi' = c' + (\sigma - u)\tan\varphi' \tag{5-4}$$

Where all strength parameters c' and φ' are expressed in terms of the effective normal stress $\sigma'(=\sigma - u)$. The concept states that soil strength is controlled by the effective stress (stresses in the soil's skeleton) rather than the total stress.

式中,强度参数c'和φ'均使有效正应力$\sigma'(=\sigma - u)$表示。土的有效应力原理认为土体强度由有效应力(土骨架的应力)控制,而不是总应力控制。

The coordinates of the tangent point τ_f and σ_f can be written with respect to the major and minor principal stress σ_1 and σ_3 of the critical stress circle, as shown in Figure 5-7.

如图5-7所示,切点τ_f和σ_f的坐标可采用极限应力圆大、小主应力σ_1和σ_3来表示。

$$\tau_f = \frac{1}{2}(\sigma_1 - \sigma_3)_f \sin 2\theta_f \tag{5-5}$$

$$\sigma_f = \frac{1}{2}(\sigma_1 + \sigma_3)_f + \frac{1}{2}(\sigma_1 - \sigma_3)_f \cos 2\theta_f \tag{5-6}$$

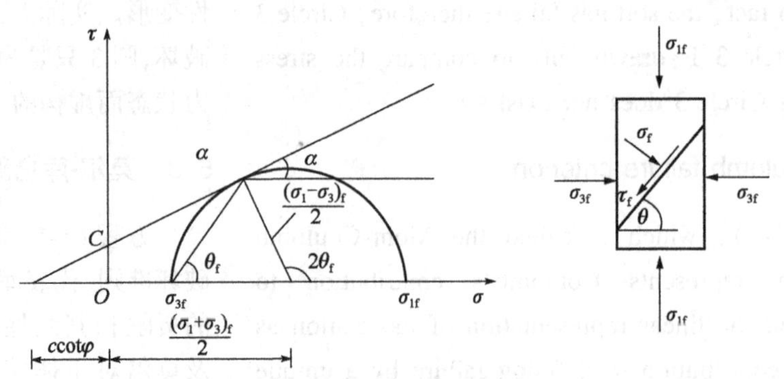

Figure 5-7 Critical stress circle

图 5-7 极限应力圆

式中,θ_f 为最大主应力面和破裂面的理论夹角,可以表示为:

Where θ_f is the theoretical angle between the major principle plane and the failure plane. It can be expressed:

$$\theta_f = 45° + \frac{\varphi}{2} \tag{5-7}$$

土体破坏时主应力与抗剪强度参数之间的关系为:

The relation between the principle stresses at failure and the shear strength parameters can be obtained:

$$\sin\varphi = \frac{\frac{1}{2}(\sigma_1 - \sigma_3)_f}{\frac{1}{2}(\sigma_1 + \sigma_3)_f + c \cdot \cot\varphi} \tag{5-8}$$

$$\frac{1}{2}(\sigma_1 - \sigma_3)_f = \frac{1}{2}(\sigma_1 + \sigma_3)_f \cdot \sin\varphi + c \cdot \cos\varphi \tag{5-9}$$

由式(5-5)和(5-6)可以得到:

Equations (5-5) and (5-6) can be rewritten as:

$$\sigma_{1f} = \sigma_{3f} \cdot \tan^2\left(45° + \frac{\varphi}{2}\right) + 2c \cdot \tan\left(45° + \frac{\varphi}{2}\right) \tag{5-10}$$

$$\sigma_{3f} = \sigma_{1f} \cdot \tan^2\left(45° - \frac{\varphi}{2}\right) - 2c \cdot \tan\left(45° - \frac{\varphi}{2}\right) \tag{5-11}$$

5.4 三轴压缩试验

5.4 Triaxial compression test

5.4.1 试验装置

5.4.1 Test setup

通常使用三轴仪来测定一般应力和排水条件下土的抗剪强度。如图 5-8 所示,对圆柱形试样施加三个主应力 σ_1、σ_2 和 σ_3,其中 $\sigma_2 = \sigma_3$。剪切过程中,侧向围压保持不变,而轴向应力持续增大至土样破坏。因此,轴

A triaxial compression test device is routinely used to determine the shear strength of soils for more general stresses and drainage conditions. It applies three principal stresses, σ_1, σ_2, and σ_3, to a cylindrical specimen; the intermediate principal stress σ_2 is equal to the minor principal stress σ_3, as seen in Figure 5-8. The axial stress is increased until failure, while the lateral confining pressure is kept constant during the shear. Thus, the axial stress is the major principal stress σ_1 and the lateral confining pressure is the minor principal stress σ_3.

向应力为最大主应力 σ_1,围压为最小主应力 σ_3。

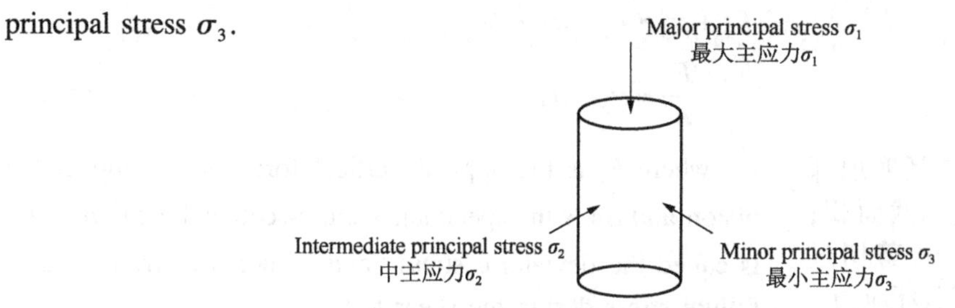

Figure 5-8 Triaxial stresses on a cylindrical specimen($\sigma_2 = \sigma_3$ in triaxial test)

图 5-8 圆柱体试样的三向应力分布(三轴试验中 $\sigma_2 = \sigma_3$)

Figure 5-9 shows the schematic setup of a typical triaxial compression test device. In this system, a specimen is enclosed in a thin rubber membrane and placed on a loading platform. The lateral confining pressure is applied through a thin rubber membrane to the specimen via chamber pressure. During the test, the confining pressure, in general, is kept constant and the axial compressive force F_v is increased to failure. The vertical deformation δ_v is measured to compute the axial strain.

图 5-9 为常规三轴试验装置,首先,将试样置于薄橡皮膜中,并将其放置在加载平台上;然后,将压力室产生的液压作用作为围压通过橡皮膜施加到试样上,试验过程中,围压一般保持恒定;最后,在试样顶端施加轴向压力 F_v,直至土样破坏,测量轴向变形 δ_v 计算轴向应变 ε_v ($=\delta_v/L_0$)。

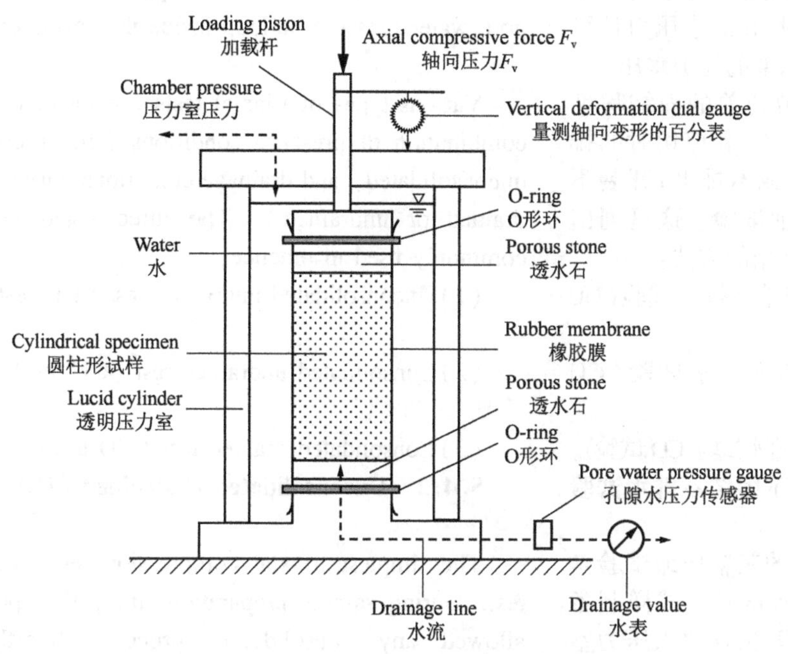

Figure 5-9 A typical triaxial test setup

图 5-9 三轴试验仪

For a free-body diagram of the upper section of soil specimen, as seen in Figure 5-9, the vertical force equilibrium

图 5-9 所示土样作为隔离体,忽略土样、荷载帽和活塞的

重量,建立轴向力平衡:

is established by neglecting the weights of the soil, loading cap, and loading piston as follows:

$$F_v + \sigma_3 \cdot A_s = \sigma_1 \cdot A_s$$

$$\frac{F_v}{A_s} = \sigma_1 - \sigma_3 \tag{5-12}$$

式中,F_v为施加在活塞顶部的垂向荷载;A_s为试样截面积;$(\sigma_1 - \sigma_3)$称为偏应力,剪切过程中由零逐渐增加至破坏应力。

where F_v is the applied vertical force on the top of the piston and As is the specimen's cross-sectional area. $\sigma_1 - \sigma_3$ is called the deviatoric stress, and it increases from zero to failure stress during the shear test.

实际上,三轴试验过程中需要注意制样和剪切方法以及剪切过程中孔压消散问题。考虑试样的制备方法及剪切过程中孔隙水压力的消散方式。如图5-9所示,排水管通过透水石与试样内部连通。试样剪切之前如果开启排水阀,使先期围压产生的孔压消散,则进行固结试验;如果关闭排水阀进行不固结试验。在试样剪切过程中,可以关闭排水阀进行不排水剪,也可以打开排水阀进行排水剪。当排水阀关闭时,孔隙水压力计可以监测试样内部的孔隙水压。

In actual practice, a triaxial test requires more detailed techniques concerning how the specimen is prepared and sheared, and pore water pressure dissipation during the shear. In this respect, the drainage line and the drainage valve shown in Figure 5-9 play significant roles. As seen in the figure, the drainage line is connected from the inside of the specimen through the porous stone. During the pre-shearing process, the drainage valve is kept open to allow the dissipation of induced pore water pressure for a consolidated test or kept closed for an unconsolidated test. During the shear, the drainage valve could be either closed for an undrained test or opened for a drained test. When the valve is closed, the pore water pressure gauge monitors the pore water pressure buildup inside the specimen.

可根据剪切前的固结程度(固结或未固结)和剪切时的排水条件(排水或不排水)开展不同类型的三轴试验。这里列出三种实践中常用的方法:

(1)不固结不排水试验(UU试验);

(2)固结不排水试验(CU试验);

(3)固结排水试验(CD试验)。

Various types of triaxial tests can be conducted using a combination of preshear conditions (either consolidated or unconsolidated) and drainage conditions during shear (either drained or undrained). The three listed here are most commonly used in practice:

(1) Unconsolidated undrained test (UU test);

(2) Consolidated undrained test (CU test);

(3) Consolidated drained test (CD test).

5.4.2 不固结不排水试验(UU)

5.4.2 Unconsolidated Undrained (UU) test

最简单的三轴压缩试验是不固结不排水试验。试样制备时无固结过程,试样放入压力室后立即施加围压,随后在不排水条件下剪切。试样制备和剪切过程中,排水阀保持关闭并在短时间内完成剪切,孔隙水不会从

The simplest triaxial compression test would be a UU test. During sample preparation time, the specimen is not allowed any consolidation process. Shortly after the specimen is placed in the chamber, the confining stress is applied through the specimen membrane and sheared under undrained conditions; that is, the drainage valve is kept closed, and it is sheared in a short time. During the sample preparation and shearing processes, there will be no escape of the pore water from the specimen; thus, no change in the water content or volume of the specimen will take place. Therefore, the anticipated soil strengths are the same for any confining stress. Figure 5-10 shows Mohr's circles at failure for similar fully saturated specimens under different confining stresses from the UU test. The diameters of Mohr's circles at failure are the same and the drawn failure envelope is horizontal, which implies $\varphi = 0$.

试样中逸出,使试样含水量始终保持不变,即试样体积不变。因此,不同围压下的土样强度均相等。图 5-10 为不同围压下饱和土样不固结不排水试验的破坏莫尔圆,各圆直径相同,破坏包络线为水平直线,即 $\varphi = 0$。

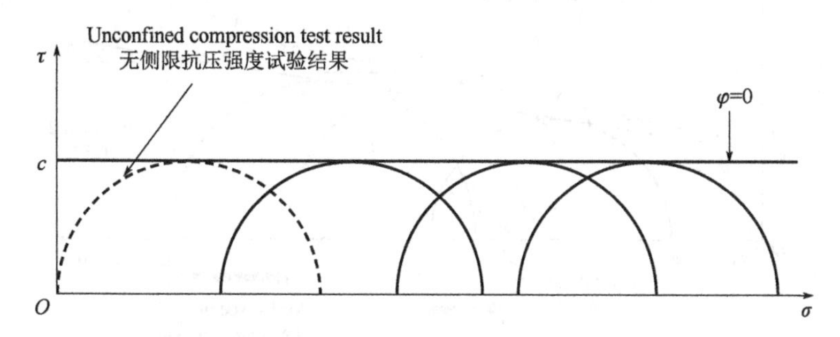

Figure 5-10 UU test result
图 5-10 UU 试验结果

5.4.3 Consolidated Undrained (CU) test

This is the most widely used triaxial shear test in practice. A specimen is first fully consolidated in the triaxial cell. Then the drainage valve is closed, and the specimen is sheared by increasing the deviatoric stress $\sigma_1 - \sigma_3$ to failure while σ_3 is held constant. The pore water pressure u is measured during the shearing process. The deviatoric stress $\sigma_1 - \sigma_3$ and pore water pressure u with respect to the vertical strain ε_1 are measured during the test.

From the values of σ_3, σ_{1f}, and u_f at failure for a given specimen, the total principal stresses (σ_3 and σ_{1f}), as well as the effective principal stresses ($\sigma'_{3f} = \sigma_3 - u_f$ and $\sigma'_{1f} = \sigma_{1f} - u_f$), are calculated, and Mohr's circles at failure are drawn for the total stress and the effective stress. These circles are shown in Figure 5-11, where solid lines indicate total stress and dotted lines indicate effective stress. The diameters of Mohr's circles are the same for both the total stress and the effective stress, but the latter circle is shifted toward the left by an amount equal to u_f for a positive pore water pressure at failure.

Similarly, another set of failure Mohr's circles are drawn for differently consolidated specimen 2. The failure

5.4.3 固结不排水试验(CU)

固结不排水试验是实践中应用最广泛的三轴剪切试验。首先,试样在压力室完全固结;然后,保持 σ_3 不变,关闭排水阀,增大偏应力 $\sigma_1 - \sigma_3$,直至试样破坏;同时,测量剪切过程中的孔隙水压力 u。试验测量值包括 $\sigma_1 - \sigma_3$、u 及垂向应变 ε_1。

根据试样破坏时的 σ_3、σ_{1f} 和 u_f,计算总应力(σ_3, σ_{1f})和有效应力($\sigma'_{3f} = \sigma_3 - u_f$, $\sigma'_{1f} = \sigma_{1f} - u_f$),并绘制破坏莫尔圆。如图 5-11 所示,实线表示总应力,虚线表示有效应力。两个莫尔圆直径相同,但试样破坏时存在正孔隙水压力 u_f,有效应力莫尔圆位置向左移动。

类似地,对不同固结度的试样 2 绘制破坏莫尔圆。绘制与莫尔圆相切的破坏包络线,确定总应力强度参数 c 和 φ 以及有效应力参数 c'和 φ',如图 5-11 所示。为了保证 c、φ、c'和 φ'值的可靠性,一般需要进行三次或三次以上不同固结压力的固结不排水试验。

envelopes, which are tangent to those circles, are then drawn to determine the total stress strength parameters c and φ as well as the effective stress parameters c' and φ', as seen in Figure 5-11. In practice, three or more CU tests with different consolidation stresses for similar specimens are performed to determine reliable c, φ, c', and φ' values.

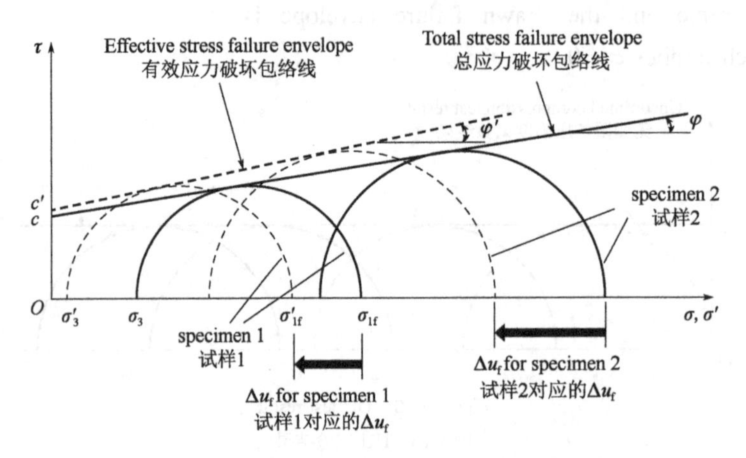

Figure 5-11 Total stress and effective stress analyses from CU test

图 5-11 CU 试验的总应力和有效应力分析

例题 5-1

图 5-12 为三个不同围压下土体的三轴试验结果,绘制轴向应变 ε_1 与偏应力$(\sigma_1-\sigma_3)$关系曲线,并确定试验的破坏强度$(\sigma_1-\sigma_3)_f$。绘制三个试样破坏时的莫尔圆,并确定黏聚力 c 和内摩擦角。

Example 5-1

Triaxial test data for similar soil samples at three different confining pressures are shown in Figure 5-12. The deviatoric stress $(\sigma_1-\sigma_3)$ is plotted with respect to the vertical strain ε_1, and the failure strengths $(\sigma_1-\sigma_3)$ $(\sigma_1-\sigma_3)_f$ are identified for those tests. After drawing Mohr's circles at failure for the three specimens, determine the cohesion component c and the angle of internal friction of the soil.

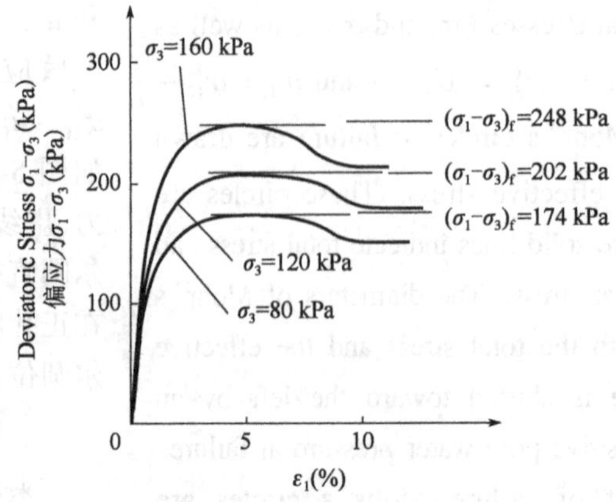

Figure 5-12 Example 5-1 problem (results from triaxial tests)

图 5-12 例题 5-1 问题(三轴试验结果)

Solution

解：

From the data, for specimen 1:

试样1有：

$$\sigma_3 = 80(\text{kPa})$$

$(\sigma_1 - \sigma_3)_f = 174$ kPa and thus:

因为$(\sigma_1 - \sigma_3)_f = 174$ kPa，可得：

$$\sigma_{1f} = (\sigma_1 - \sigma_3)_f + \sigma_3 = 174 + 80 = 254(\text{kPa})$$

From the data, for specimen 2:

试样2有：

$$\sigma_3 = 120(\text{kPa})$$

$(\sigma_1 - \sigma_3)_f = 202$ kPa and thus:

因为$(\sigma_1 - \sigma_3)_f = 202$ kPa，可得：

$$\sigma_{1f} = (\sigma_1 - \sigma_3)_f + \sigma_3 = 202 + 120 = 322(\text{kPa})$$

From the data, for specimen 3:

试样3：

$$\sigma_3 = 160(\text{kPa})$$

$(\sigma_1 - \sigma_3)_f = 248$ kPa and thus:

因为$(\sigma_1 - \sigma_3)_f = 248$ kPa，可得：

$$\sigma_{1f} = (\sigma_1 - \sigma_3)_f + \sigma_3 = 248 + 160 = 408(\text{kPa})$$

Based on the preceding σ_{1f} and σ_3 values, Mohr's circles at failure are constructed in Figure 5-13. A failure envelope is also drawn by just touching these Mohr's circles at failure, and the cohesion c and the angle of internal friction φ are read as 36 kPa and 18.5°, respectively, as seen.

基于上述σ_{1f}和σ_3值，绘制试样破坏时的莫尔圆，如图5-13所示。绘制破坏包络线，黏聚力和内摩擦角分别为36kPa和18.5°。

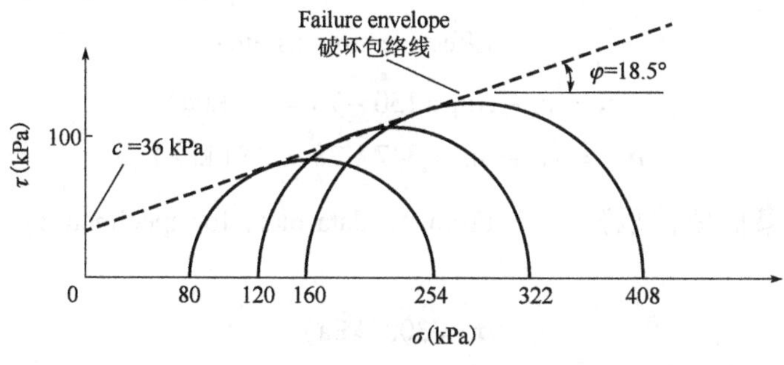

Figure 5-13 Example 5-1 (determination of c and φ)

图5-13 例题5-1(c和φ的测定)

Example 5-2

例题5-2

Consolidated undrained triaxial tests for two similar specimens with different consolidation stresses were performed. The data obtained, shown in Figure 5-14, include pore water pressure measurements. Plot Mohr's circles at failure for two specimens in terms of both the total

对两个固结应力不同的试样进行固结不排水三轴试验，试验结果如图5-14所示，虚线表示孔隙水压力。绘制两个试样破坏时总应力和有效应力对应的莫尔圆,并根据总应力破坏包络线和有效应力破坏包络线确定抗剪强度参数 c、φ、c'、φ'。

stress and the effective stress, and determine the shear strength parameters c and φ from the total stress failure envelope and c' and φ' from the effective stress failure envelope.

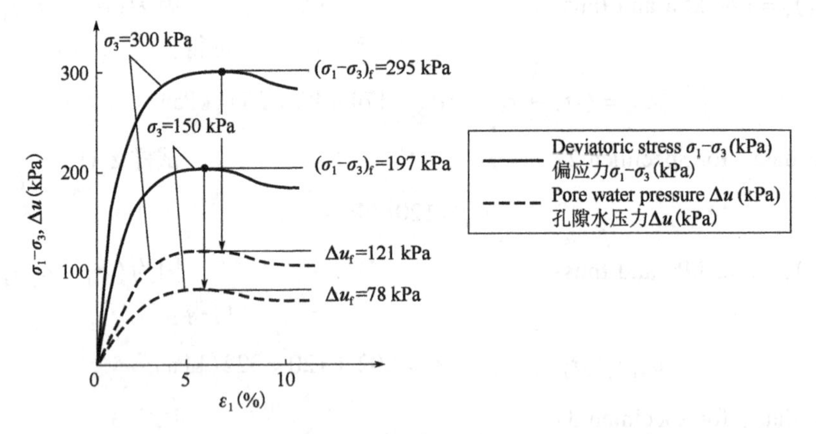

Figure 5-14 Example 5-2 problem (results from CU tests)

图 5-14 例题 5-2 问题(CU 测试结果)

解:

根据试验结果曲线,试样1有:

Solution

From the data plot, for specimen 1:

$$\sigma_3 = 150(\text{kPa})$$
$$(\sigma_1 - \sigma_3)_f = 197(\text{kPa})$$

因为 $\Delta u_f = 78$ kPa,所以有:

$\Delta u_f = 78$ kPa and thus:

$$\sigma_{1f} = (\sigma_1 - \sigma_3)_f + \sigma_3 = 197 + 150 = 347(\text{kPa})$$

有效应力为:

Effective stresses are:

$$\sigma'_3 = \sigma_3 - \Delta u_f = 150 - 78 = 72(\text{kPa})$$
$$\sigma'_{1f} = \sigma_{1f} - \Delta u_f = 347 - 78 = 269(\text{kPa})$$

根据试验结果曲线,试样2有:

From the data plot, for specimen 2:

$$\sigma_3 = 300(\text{kPa})$$
$$(\sigma_1 - \sigma_3)_f = 295(\text{kPa})$$

因为 $\Delta u_f = 121$kPa,所以有:

$\Delta u_f = 121$kPa and thus:

$$\sigma_{1f} = (\sigma_1 - \sigma_3)_f + \sigma_3 = 295 + 300 = 595(\text{kPa})$$

有效应力为:

Effective stresses are:

$$\sigma'_3 = \sigma_3 - \Delta u_f = 300 - 121 = 179(\text{kPa})$$
$$\sigma'_{1f} = \sigma_{1f} - \Delta u_f = 595 - 121 = 474(\text{kPa})$$

From these values, Mohr's circles at the failure are drawn in Figure 5-15. Accordingly, c = 42 kPa, φ = 149 for the total stress, and c' = 53 kPa, φ' = 18°, for the effective stress, are read from the plot.

根据上述试验结果,绘制试样破坏时的莫尔圆,如图5-15所示。相应地,总应力抗剪强度参数 c = 42 kPa, φ = 14°,有效应力抗剪强度参数 c' = 53 kPa, φ' = 18°。

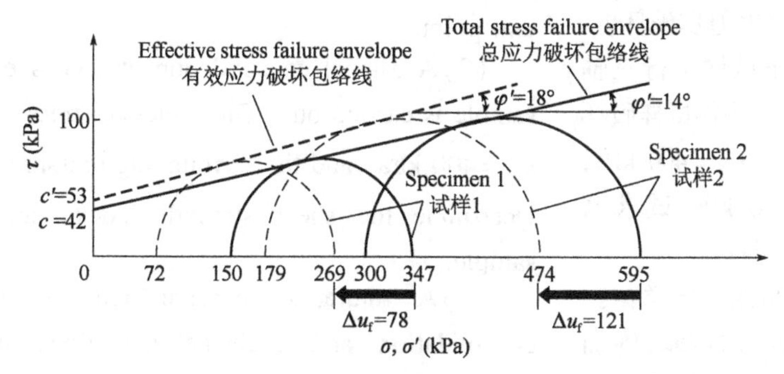

Figure 5-15 Example 5-2 (determination of c, φ and c', φ')

图5-15 例题5-2(c, φ 和 c', φ' 的测定)

5.4.4 Consolidated Drained (CD) triaxial test

In this type of test, the specimen is first fully consolidated, and then it is sheared slowly to allow the generated pore water pressure to be fully dissipated. Let us assume that a clay specimen is present in a mud state and then a consolidated drained test is performed. For nearly zero consolidation pressure, the strength of the specimen is nearly equal to zero since the specimen's initial water content is above the liquid limit. With a small consolidation pressure, the specimen gains some strength due to the consolidation process. During shear, it also gains some more strength due to the drainage of water. With higher consolidation pressure, it gains more strength due to the continued reduction in water content through the processes of consolidation and drained shear. Accordingly, Mohr's failure circles that are nearly proportional in size to their consolidation pressures are drawn to define the failure envelope of the soil, as seen in Figure 5-16.

5.4.4 固结排水试验(CD)

对于固结排水试验,首先对试样进行排水固结;然后在排水条件下对试样进行缓慢剪切直至试样破坏,以使孔隙水压力完全消散。假设一黏土试样呈泥浆状态,如果固结时围压为零,那么该试样的强度几乎等于零;在较小的固结压力下,试样可以获得一定的强度。在剪切过程中,由于排水,试样强度会进一步增大。固结压力越大,固结和排水剪切过程中试样的含水量降低幅度越大,其强度增大越显著。如图5-16所示绘制与固结压力成比例的破坏莫尔圆和破坏包络线。

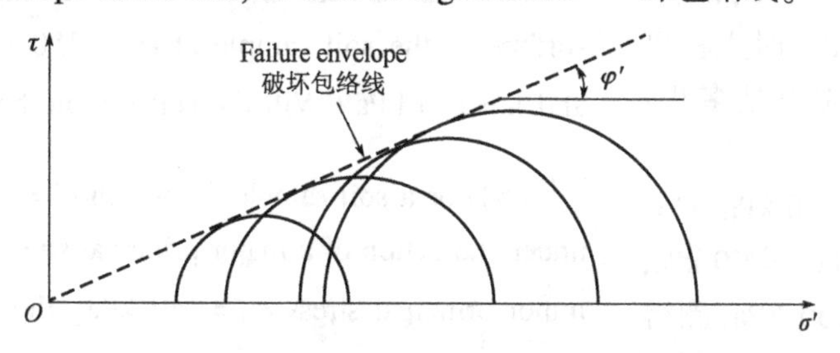

Figure 5-16 Failure envelope from CD test for soils

图5-16 土的固结排水试验的破坏包络线

习　题

(1)已知地基土的抗剪强度指标 $c=10$ kPa,$\varphi=30°$,问当地基中某点的最大主应力 $\sigma_1=400$ kPa,而最小主应力 σ_3 为多少时,该点刚好发生剪切破坏?

(2)对某砂土试样进行三轴固结排水剪切试验,测得试样破坏时的主应力差 $\sigma_1-\sigma_3=400$ kPa,周围压力 $\sigma_3=100$ kPa,试求该砂土的抗剪强度指标。

(3)一饱和黏性土试样在三轴仪中进行固结排水试验,施加围压 $\sigma_3=200$ kPa,试样破坏时的主应力差 $\sigma_1-\sigma_3=300$ kPa,整理试验成果得有效应力强度指标 $c'=75.1$ kPa,$\varphi'=30°$。问:①破裂面上的法向应力和剪应力以及试样中的最大剪应力为多少?②为什么试样的破坏发生在与最大主应力成 $\alpha=60°$ 的平面而不发生在最大剪应力的作用面?

(4)某饱和土样,已知土的抗剪强度指标为 $c_u=35$ kPa,$\varphi_u=0$,$c_{cu}=12$ kPa,$\varphi_{cu}=12°$,$c'=3$ kPa,$\varphi'=28°$,则:

①若该土样在 $\sigma_3=200$ kPa 作用下进行三轴固结不排水剪切试验,则破坏时的 σ_1 约为多少?

②在 $\sigma_3=250$ kPa,$\sigma_1=400$ kPa,$u=160$ kPa 时土样可能破裂面上的剪应力是多少?土样是否会破坏?

(5)某土样 $c'=20$ kPa,$\varphi'=30°$,承受最大主应力 $\sigma_1=420$ kPa,最小主应力 $\sigma_3=150$ kPa,测得孔隙水压力 $u=46$ kPa,试判断土样是否达到极限平衡状态。

Excercises

(1) The shear strength index values of a foundation soil are $c=10$ kPa and $\varphi=30°$, and the major principal stress at a certain point in the foundation is $\sigma_1=400$ kPa. Calculate the minor principal stress when the shear failure occurs at that point.

(2) A triaxial consolidation drained shear test of a sand sample is carried out. The principal stress difference is $\sigma_1-\sigma_3=400$ kPa, and the surrounding pressure is $\sigma_3=100$ kPa. Determine the shear strength index values of the sand sample.

(3) A saturated cohesive soil sample is tested in consolidation and drainage conditions using a triaxial apparatus. When a surrounding pressure of $\sigma_3=200$ kPa is applied, the principal stress difference is $\sigma_1-\sigma_3=200$ kPa when the sample is damaged. The effective shear strength index values are found to be $c'=75.1$ kPa and $\varphi'=30°$ from the test results. ① Determine the normal and shear stresses on the failure surface and the maximum shear stress in the specimen. ② Why does the failure surface of the specimen occur in the plane at $\alpha=60°$ to the maximum principal stress and not in the action surface of the maximum shear stress?

(4) For a fully saturated soil sample, the shear strength index values of a soil are known to be $c_u=35$ kPa, $\varphi_u=0$; $c_{cu}=12$ kPa, $\varphi_{cu}=12°$; and $c'=3$ kPa, $\varphi'=28°$.

①If the soil sample is subjected to $\sigma_3=250$ kPa in a triaxial consolidated undrained shear test, what is σ_1 at failure?

② What is the shear stress on the possible fracture surface of the soil sample at $\sigma_3=250$ kPa, $\sigma_1=400$ kPa, and $u=160$ kPa? Will the soil sample be destroyed?

(5) For a soil sample for which $c'=20$ kPa and $\varphi'=30°$, under the action of a major principal stress $\sigma_1=420$ kPa and a minor principal stress $\sigma_3=150$ kPa, the pore water pressure of the soil sample is measured to be $u=46$ kPa. Judge whether the soil sample reaches the limit equilibrium state.

Lateral Earth Pressure 6 Chapter 土压力

6.1 Introduction

Earth retaining structures are commonly used to support soils and structures and thus maintain a difference in elevation of the ground surface. Earth pressure is the lateral force that acts between retaining structures and the soil masses being retained. Proper estimation of lateral earth pressure against such structures is critical to their adequate design and performance.

Figure 6-1 shows the schematic of a vertical underground wall in a soil mass. The amount of lateral earth pressure against the wall depends on how the wall moves relative to the soil mass. If the wall does not move at all, then the same lateral pressure is anticipated on the left and right faces of the wall. However, when the wall is moved toward the right, the wall pushes against the soil mass; thus, a higher lateral pressure is developed on the right face of the wall. In contrast the lateral pressure is reduced on the left face of the wall since the wall moves away from the soil.

It is convenient to express the lateral earth pressure σ_h in terms of its ratio to the vertical stress σ_v, as follows:

$$\sigma_h = K\sigma_v \tag{6-1}$$

Where K is the coefficient of lateral earth pressure and changes depending on the wall movement relative to the soil mass.

6.1 概述

挡土结构是用于支挡土质边坡和结构以保持地表高度差。土压力是作用于挡土结构和土体之间的侧向力。对于挡土结构的设计与使用,土压力计算极为关键。

图 6-1 为嵌入土体的竖直地下墙。墙体承受的侧向土压力大小取决于墙体相对于土体的移动方式。如果墙体不移动,则在墙体左侧和右侧面上的土压力近似相等。当墙体向右侧移动时,墙体会对右侧土体产生推力,从而在墙体的右侧面产生较高的侧向土压力。同时,由于墙体逐渐远离左侧土体,其左侧面的侧向土压力会减小。

通常将侧向土压力 σ_h表示为其与竖向应力 σ_v的比值:

式中,K 为土压力系数,其变化取决于墙体与土体的相对移动方向。

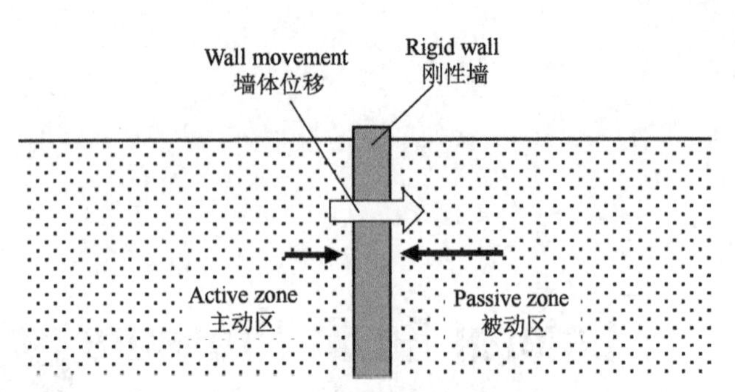

Figure 6-1 Lateral earth pressure against an underground wall

图 6-1 地下墙的侧向土压力

图 6-2 为 K 值随墙体移动的变化规律。当墙体向右侧移动时,K 增加,并在墙体充分移动的情况下达到最大值 K_p。此时,墙体右侧的土体发生破坏,这一临界侧向土压力称为被动土压力,K_p 称为被动土压力系数。而在墙体左侧,K 值减小并接近最小值 K_a,此时的土压力称为主动土压力,K_a 称为主动土压力系数。当墙身位移为零时,K 值为 K_0,称为静止土压力系数。

Figure 6-2 shows a plot of the change in K with the amount of wall movement. When the wall is moved toward the right, K increases and reaches a maximum value K_p with sufficient wall movement. At this stage, the soil mass on the right side of the wall fails. The lateral earth pressure at this critical stage is called passive earth pressure, and K_p is called the coefficient of passive earth pressure. On the left side of the soil mass, the K value decreases and approaches the minimum value K_a. This critical pressure is called active earth pressure, and K_a is called the coefficient of active earth pressure. With zero wall movement, the K value is K_0, which is called the coefficient of lateral earth pressure at rest.

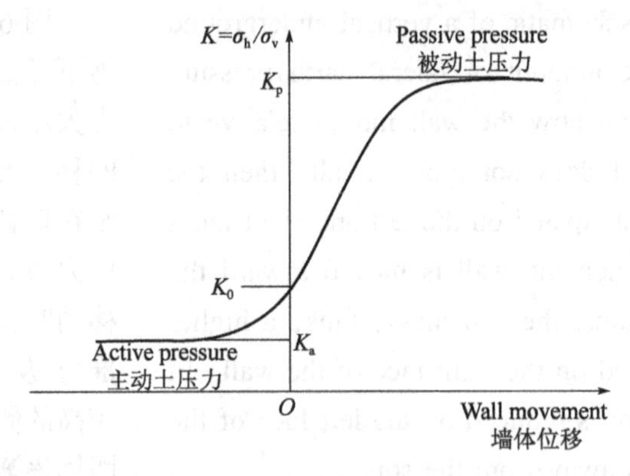

Figure 6-2 Coefficient of lateral earth pressure K versus wall movement

图 6-2 土压力系数 K 与墙体位移的关系

根据图 6-2 可知以下关系:

As shown in Figure 6-2, the following relationship can be observed:

$$K_p > K_0 > K_a$$

静止土压力是指当墙体没有移动时的侧向土压力。从图 6-2 可以看到,土压力系数 K 在墙体

The at-rest pressure is the lateral earth pressure when there is no wall movement. As can be seen in Figure 6-2, the coefficient of lateral earth pressure K changes sharply in the vicinity of zero wall deformation ($K = K_0$). This implies that K_0 is very sensitive to small wall movements. There are several solutions to the problem of determining K_0.

零位移($K = K_0$)附近时急剧变化,从而表明K_0对微小的墙体移动非常敏感。以下为确定K_0的方法:

1) Elastic solution

1)弹性解

This solution is based on an assumption that soil is an elastic medium. This is a reasonable assumption since there is no wall movement in this situation:

该方法假设土体为弹性介质,在这种情况下墙体没有移动:

$$K_0 = \frac{\mu}{1-\mu} \tag{6-2}$$

Where μ is the soil's Poisson's ratio. If $\mu = 0.3$ is assumed for sands, $K_0 = 0.43$, and if $\mu = 0.4$ is assumed for clays, $K_0 = 0.67$.

式中,μ为土的泊松比。对于砂土,假定$\mu = 0.3$,则K_0为0.43;对于黏性土,如果$\mu = 0.4$,则K_0为0.67。

2) Empirical formula

2)经验公式

Jaky (1944) developed an empirical formula for K_0 for normally compacted sand:

Jaky(1944)针对正常压实的砂土提出一个经验公式:

$$K_0 = 1 - \sin\varphi' \tag{6-3}$$

Where φ' is the drained angle of internal friction of the soil. Jaky's formula is widely used because of its simplicity and its validity for loose sand.

式中,φ'为土的有效内摩擦角。Jaky公式因其简明性和对松散砂土的有效性而被广泛应用。

In a semi-infinite plane without any lateral deformation, if the top surface is horizontal, then any horizontal plane and vertical plane are the main planes. When a smooth rigid vertical wall is used to replace the soil on the left side of the vertical plane, the lateral earth pressure on the vertical wall is equivalent to the lateral pressure on the vertical plane, as shown in Figure 6-3. According to Equation (6-1), the static earth pressure has a triangular distribution with depth. For a wall with a height of H, the static earth pressure is given by:

在无侧向变形的平面半无限体中,若其顶面水平,则任一水平面和铅垂面均为主平面。现用一竖直光滑的刚性墙背取代此铅垂面左侧的土体,则此竖直墙背上的侧向压力相当于该竖直面上的侧向压力,如图6-3所示。由公式(6-1)可知,静止土压力强度沿深度呈三角形分布,对高为H的墙背,其静止土压力值为:

$$E_0 = \frac{1}{2}K_0 \cdot \gamma H^2 \tag{6-4}$$

Where γ is the bulk density of the soil. The direction of E_0 is horizontal, and the distance between the action point and the bottom of the wall is $H/3$.

式中,γ为土的重度;E_0的方向水平,作用点距墙底部为$H/3$。

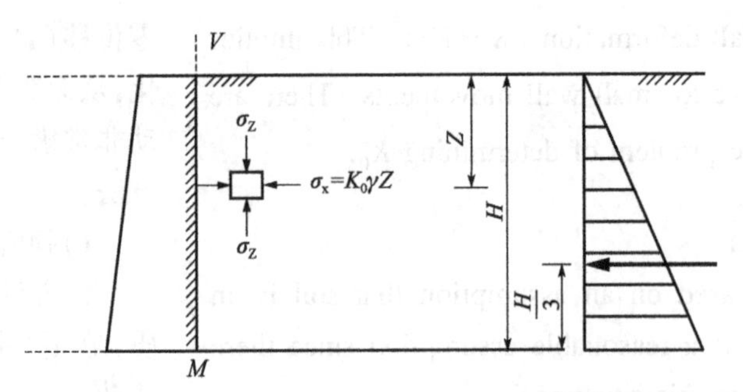

Figure 6-3 Lateral earth pressure against basement wall

图 6-3 挡土墙的土压力分布

例题 6-1

挡土墙建在基岩上,墙高 $H=8.0$ m,墙后回填中密砂,其重度和内摩擦角分别为18.5 kN/m³和30°,计算作用在墙体上的土压力。

解:

这种情况下,挡土墙建于基岩上,不会移动或旋转,土压力可按静止土压力计算。

Example 6-1

A retaining wall rests on bedrock. The height of the wall is $H=8.0$ m. The backfill of the wall is medium dense sand, and its unit weight and internal friction angle are $\gamma=18.5\ \text{kN/m}^3$ and $\varphi'=30°$. Calculate the earth pressure of the wall.

Solution

In this case, the retaining wall rests on bedrock and cannot move or rotate; thus, the earth pressure can be treated as the earth pressure at rest.

$$K_0=1-\sin\varphi'=0.5$$

$$E_0=\frac{1}{2}\gamma H^2K_0=296(\text{kN/m})$$

6.2 朗肯土压力理论

朗肯于1857年提出了墙后土体侧向土压力理论,假设墙体远离填土(主动土压力)或朝向填土(被动土压力)移动时,墙后土体会进入极限平衡状态。

如图6-4所示,刚性挡土墙支挡着墙背后的水平填土,当墙体向左侧移动足够距离导致填土破坏时,填土中的所有土单元进入极限平衡状态。由于水平填土层的垂直面和水平面上没有剪切应力,因此深度 z 处土单元的 $\sigma_v(=\gamma z)$ 和 σ_h 是主应力,而水平应力 σ_h 是该极限平衡条件下的主动土压力。

6.2 Rankine's lateral earth pressure theory

Rankine (1857) developed a theory for the lateral earth pressure behind a retaining wall. He assumed that the soil mass behind the wall enters a limit equilibrium state when a sufficient wall boundary is moved away from the backfill (in the active case) or toward the backfill (in the passive case).

A rigid wall supports the horizontal backfill, as seen in Figure 6-4. When the wall moves a sufficient distance to the left to cause a failure of backfill soil, all soil elements in the backfill enter into limit equilibrium state. Since there are no shear stresses on the vertical and horizontal planes under the horizontal backfill surface, $\sigma_v(=\gamma z)$ and σ_h at an element at a depth z are the principal stresses, and the horizontal stress σ_h is the active lateral earth pressure for this limit equilibrium state.

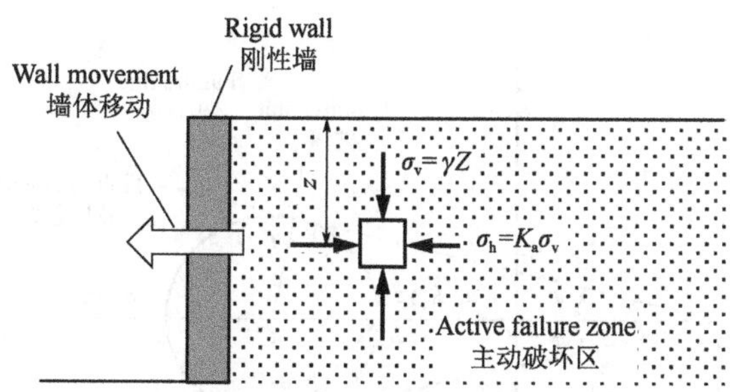

Figure 6-4 Rankine's active earth pressure development

图6-4 朗肯主动土压力

When the boundary moves against the soil mass, as seen in Figure 6-5, a higher lateral pressure is developed, and the limit equilibrium state corresponds to the passive earth pressure case.

当墙体朝向填土移动时,会产生较高的侧向压力,最终的极限平衡阶段为被动土压力工况,如图6-5所示。

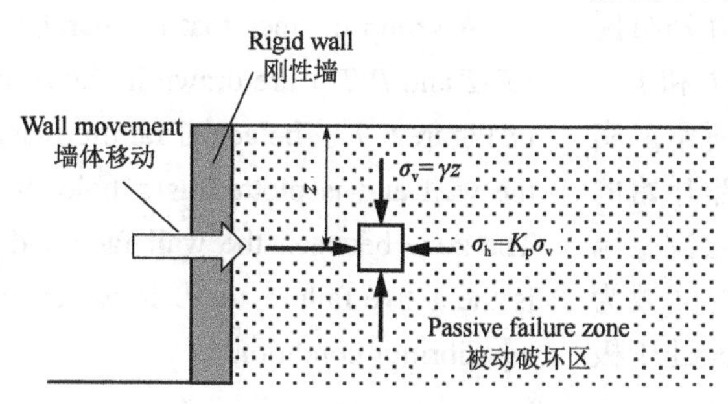

Figure 6-5 Rankine's passive earth pressure development

图6-5 朗肯被动土压力

6.2.1 Rankine active earth pressure

In the active case, σ_v is larger than σ_h, and thus, $\sigma_1 = \sigma_v$ and $\sigma_3 = \sigma_h = \sigma_a$. These are computed as follows:

6.2.1 朗肯主动土压力

在主动土压力状态下,σ_v大于σ_h,因此$\sigma_1 = \sigma_v$,$\sigma_3 = \sigma_h = \sigma_a$,$p_a$为主动土压力,其计算方法为:

$$\sigma_v = \gamma z = \sigma_1 \tag{6-5}$$

$$\sigma_h = p_a = K_a\sigma_v = K_a\gamma z = \sigma_3 \tag{6-6}$$

Equations (6-5) and (6-6) can be plotted in Mohr's circle at failure. Figure 6-6 shows a plot for the preceding situation. In the figure, take the stress circles made by σ_1 and σ_3 as the ultimate stress circle, and mark the position of the active earth pressure on the horizontal axis as point P. The failure envelope ($\tau = c + \tan\varphi$) touches the circle at point T, which is the failure point on Mohr's circle. The direction of the failure plane is determined by connecting the point P and the failure point T. Thus, the line P-T is the direction of the failure plane in the soil mass, and the line P-T' in the negative shear stress domain of the figure is also the direction of the failure plane.

式(6-5)和式(6-6)可以绘制在莫尔圆中,如图6-6所示。以σ_1、σ_3所作的应力圆为极限应力圆,将坐标轴上主动土压力所在位置记为点P。破坏包络线($\tau = c + \tan\varphi$)在T点与莫尔圆相切,该点是莫尔圆上的屈服点。P-T和P-T'为主动极限状态时两破坏面的方向。

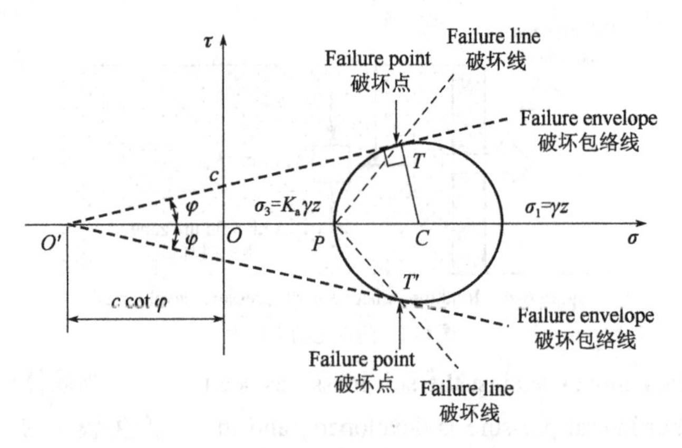

Figure 6-6 Mohr's circle at active failures of soil mass

图6-6 土体处于朗肯主动状态的莫尔应力圆

在土体主动土压力分布区绘制与两条破坏线 $P-T$ 和 $P-T'$平行的线,如图6-7所示。实际的破坏面是从墙底部开始的平面,用一条黑色虚线绘制。墙面与实际破坏面之间的区域为主动破坏区,所有土单元处于极限平衡状态。

A group of lines that are parallel to the two failure lines (P-T and P-T') are drawn in the active zone of the soil mass in Figure 6-7. The actual failure plane starts from the base of the wall and is plotted as a bold dashed line in the figure. The zone between the wall face and the actual failure plane is the active failure zone, in which all elements are in a limit equilibrium condition.

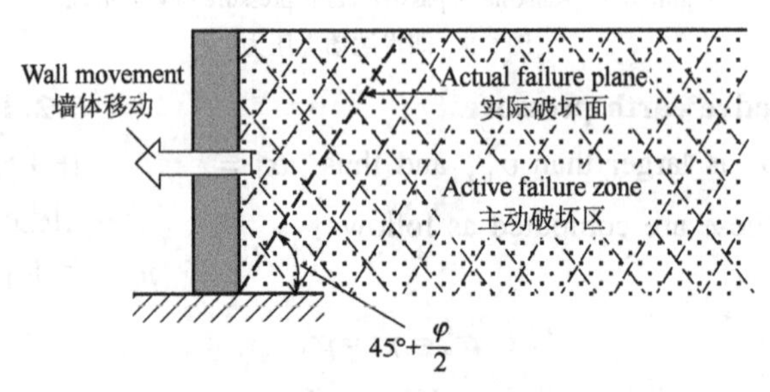

Figure 6-7 Potential active failure lines in soil mass

图6-7 土体的潜在主动破坏线

根据式(6-6),可知 $\sigma_1=\gamma z$,以及:

From Equation (6-6), we know that $\sigma_1 = \gamma z$, and:

$$\sigma_3 = p_a = \gamma z \frac{1-\sin\varphi}{1+\sin\varphi} - 2c\frac{\cos\varphi}{1+\sin\varphi} = \gamma z \tan^2\left(45° - \frac{\varphi}{2}\right) - 2c\tan\left(45° - \frac{\varphi}{2}\right) \tag{6-7}$$

此时存在两种情况:

We have two cases.

1) $c=0$(粗粒土)

1) $c = 0$ (coarse-grained soils)

式(6-7)变为:

Equation (6-7) becomes:

$$p_a = \gamma z \tan^2\left(45° - \frac{\varphi}{2}\right) = \gamma z K_a \tag{6-8}$$

Where:

式中:

$$K_a = \tan^2\left(45° - \frac{\varphi}{2}\right) \tag{6-9}$$

The Rankine active lateral earth pressure p_a acts normal to the vertical wall and increases linearly with depth z and with a slope of $1/(\gamma K_a)$, as seen in Figure 6-8. The total resultant active thrust E_a is given by:

朗肯主动土压力 p_a 垂直作用于竖直墙背，并随深度 z 线性增加，呈三角形分布，斜率为 $1/(\gamma K_a)$，如图 6-8 所示。此时墙背所受的总主动土压力 E_a 为：

$$E_a = \frac{1}{2} K_a \gamma H^2 \tag{6-10}$$

E_a is applied at the $H/3$ point from the base of the wall.

E_a 作用在距墙底 $H/3$ 处。

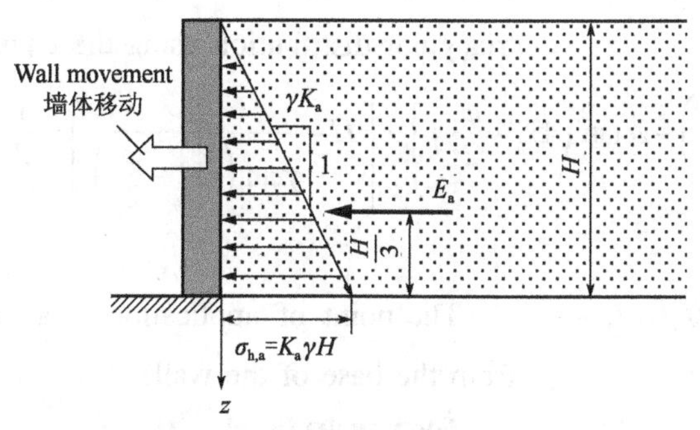

Figure 6-8 Rankine's active earth pressure distribution for $c = 0$

图 6-8 朗肯主动土压力分布($c = 0$)

2) $c > 0$

Equation (6-7) indicates that p_a increases linearly with increasing depth z. However, p_a can take a non-zero negative value at the ground surface (at $z = 0$), and thus, a negative pressure zone (tension zone) can exist near the ground surface. The distribution of p_a is plotted in Figure 6-9.

2) $c > 0$

式(6-7)表示 p_a 随深度 z 的增加而线性增大。但是，p_a 在地表($z = 0$ 时)为负值，即在地表附近存在一个负压区(张拉区)。p_a 分布规律如图 6-9 所示。

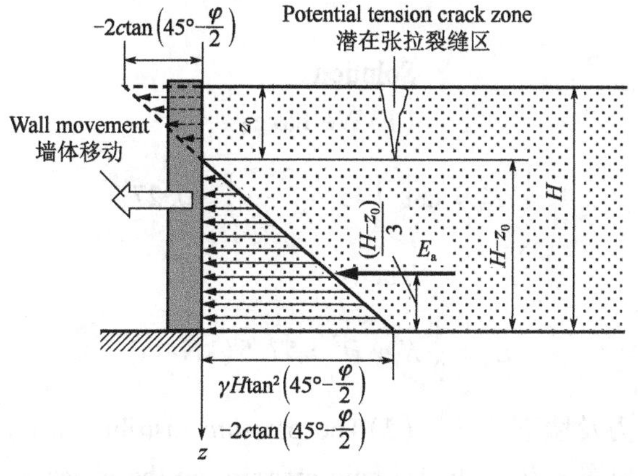

Figure 6-9 Rankine's active earth pressure distribution for $c > 0$

图 6-9 朗肯主动土压力分布($c > 0$)

张拉区深度 z_0 可通过令式(6-7)等于零求得：

The tension zone depth z_0 can be obtained by setting Equation (6-7) to zero. Thus, we obtain:

$$z_0 = \frac{2c}{\gamma \tan\left(45° - \frac{\varphi}{2}\right)} = \frac{2c}{\gamma \sqrt{K_a}} \tag{6-11}$$

由于墙背与填土交界面无法承受拉力，即张拉区的侧向应力为零，因此仅考虑 z_0 到 H 深度之间的压力三角形，如图 6-9 所示。从 $z=0$ 到 z_0 的受拉区易出现裂缝，称为拉裂区。总主动土压力 E_a 可根据三角形面积计算：

Since the interface between the wall and the soil cannot sustain tension in most cases, the lateral stress at the tension zone is taken as zero, and thus, a linear pressure distribution starts at z_0 with respect to the wall height H, as seen in Figure 6-9. The zone from $z = 0$ to z_0 can potentially cause tension cracks in the ground and is called a tension crack zone. The total active thrust E_a can be calculated from a triangular distribution using the equation:

$$E_a = \frac{1}{2}\left[\gamma H \tan^2\left(45° - \frac{\varphi}{2}\right) - 2c\tan\left(45° - \frac{\varphi}{2}\right)\right] \cdot \left[H - \frac{2c}{\gamma \tan\left(45° - \frac{\varphi}{2}\right)}\right] = \frac{1}{2}K_a \gamma H^2 - 2\sqrt{K_a}cH + 2\frac{c^2}{\gamma} \tag{6-12}$$

作用点距墙底的高度为$(H-z_0)/3$。

The point of application is at a height of $(H - z_0)/3$ from the base of the wall.

例题 6-2

(1)计算一个 5m 高竖直挡土墙(墙后填土为砂土，重度 17 kN/m^3)的主动土压力，$\varphi' = 35°$；填土顶部表面水平，地下水位低于墙底。

(2)如果地下水位上升到填土顶面以下 2 m 处，计算墙背上的土压力，砂土的饱和重度为 20 kN/m^3。

Example 6-2

(1) Calculate the total active thrust on a 5m high vertical wall that retains sand with a unit weight of 17 kN/m^3 and $\varphi' = 35°$. The surface of the sand is horizontal, and the water table is below the bottom of the wall.

(2) Determine the thrust on the wall if the water table rises to a level 2 m below the surface of the sand. The saturated unit weight of the sand is 20 kN/m^3.

解：

(1)已知：

Solution

(1) We have:

$$K_a = \tan^2\left(45° - \frac{35°}{2}\right) = 0.27$$

则有：

Then:

$$E_a = \frac{1}{2}K_a \gamma H^2 = 57.5 (\text{kN/m})$$

(2)墙背上的土压力及墙下部 3 m 处的静水压力如图 6-10 所示。

(2) The pressure distribution on the wall, including the hydrostatic pressure on the lower 3 m of the wall, is shown in Figure 6-10.

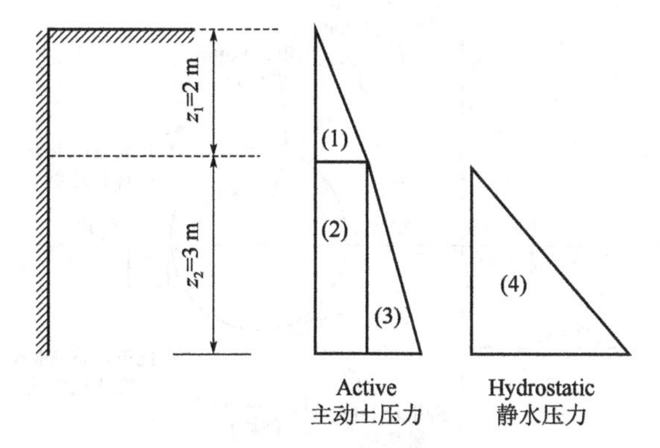

Figure 6-10 Diagram for Example 6-2
图 6-10 例题 6-2 的附图

Above the water table, the active earth pressure is:

水位线以上的主动土压力为：

$$E_{a1} = K_a \gamma z_1 = 0.27 \times 17 \times 2 = 9.18(\text{kPa})$$

Below the water table, the active earth pressure should be calculated in terms of the effective weight of the soil as follows:

水位线以下的主动土压力按土的有效重度计算，故墙底主动土压力为：

$$E_{a2} = E_{a1} + K_a \gamma' z_2 = 9.18 + 0.27 \times (20 - 9.8) \times 3 = 17.44(\text{kPa})$$

The water pressure is:

水压力为：

$$P_w = \gamma_w z_2 = 9.8 \times 3 = 29.4(\text{kPa})$$

Therefore, the total thrust (see Figure 6-10) is:

因此，总土压力（见图 6-10）为：

$$E_a = \frac{1}{2} \times 9.18 \times 2 + \frac{1}{2} \times (9.18 + 17.44) \times 3 + \frac{1}{2} \times 29.4 \times 3 = 93.21(\text{kN/m})$$

6.2.2 Rankine passive earth pressure

6.2.2 朗肯被动土压力

In the passive case, the lateral stress is larger than the vertical stress, and thus:

在被动极限平衡状态下，水平应力大于竖直应力，因此：

$$\sigma_v = \gamma z = \sigma_3 \tag{6-13}$$

$$\sigma_h = p_p = K_p \sigma_v = K_p \gamma z = \sigma_1 \tag{6-14}$$

Mohr's circle at failure is drawn in Figure 6-11 for the preceding passive stress condition.

朗肯被动土压力的极限莫尔应力圆如图 6-11 所示。

The directions of the passive failure lines in Figure 6-11 collectively form a group of potential passive failure lines behind the wall, as seen in Figure 6-12, and the actual passive failure plane is shown as a bold dashed line.

根据图 6-11 中被动极限状态时两个破坏面方向，绘制墙背填土潜在破坏面，如图 6-12 所示。实际破坏面用加粗黑色虚线表示。

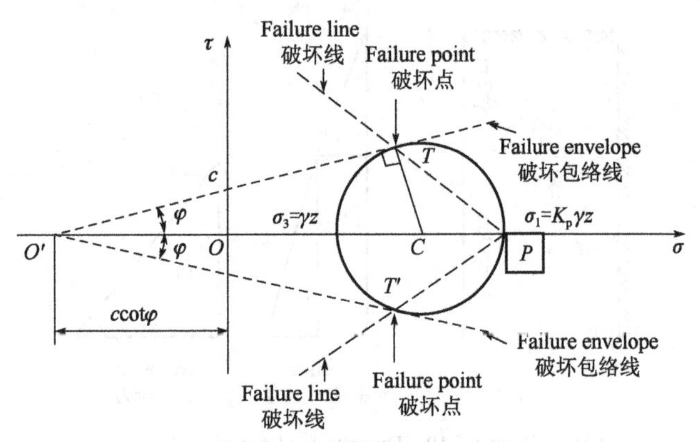

Figure 6-11 Mohr's circle for passive failures in soil mass

图 6-11 土体被动极限平衡时的莫尔应力圆

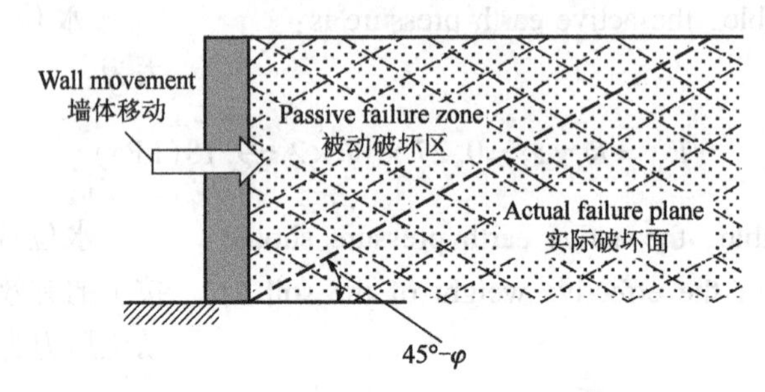

Figure 6-12 Potential passive failure lines in soil mass

图 6-12 土体潜在的被动破裂面

根据 $\sigma_3 = \gamma z$，以及：

According to $\sigma_3 = \gamma z$ and the value of σ_1:

$$\sigma_1 = p_p = \gamma z \frac{1+\sin\varphi}{1-\sin\varphi} + 2c\frac{\cos\varphi}{1-\sin\varphi} = \gamma z \tan^2\left(45° + \frac{\varphi}{2}\right) + 2c\tan\left(45° + \frac{\varphi}{2}\right) \tag{6-15}$$

此时有两种情况：

We have two cases.

1) $c = 0$(粗粒土)

1) $c = 0$ (coarse-grained soils)

式(6-15)转换为：

Equation (6-15) becomes:

$$p_p = \gamma z \tan^2\left(45° + \frac{\varphi}{2}\right) = \gamma z K_p \tag{6-16}$$

式中：

Where:

$$K_p = \tan^2\left(45° + \frac{\varphi}{2}\right) \tag{6-17}$$

从式(6-9)和(6-17)可知 $K_p = 1/K_a$。

Note that from Equations (6-9) and (6-17), the relationship $K_p = 1/K_a$ is obtained.

朗肯被动土压力 p_p 垂直作用于竖直墙背，并随深度 z 线性增加，呈三角形分布，斜率为 $1/(\gamma K_p)$，如图 6-13 所示。总被动土压力 E_p 为：

Rankine's passive lateral earth pressure p_p acts normal to the vertical wall and increases linearly with depth z with a slope of $1/(\gamma K_p)$, as seen in Figure 6-13. The resultant passive thrust E_p is given by:

$$E_p = \frac{1}{2} K_p \gamma H^2 \tag{6-18}$$

And E_p is applied at a height of $H/3$ from the base of the wall.

E_p作用在距墙底 $H/3$ 处。

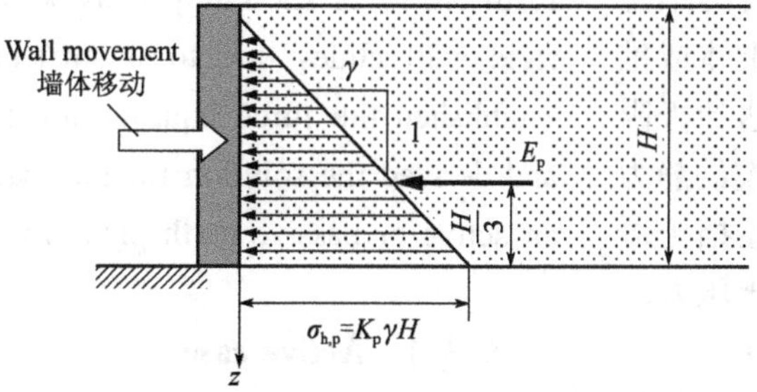

Figure 6-13 Rankine's passive earth pressure distribution($c=0$)

图 6-13 朗肯被动土压力分布($c=0$)

2) $c>0$

Equation (6-19) shows that p_p increases linearly with increasing depth z. In the passive case, there is a positive pressure at $z=0$, and there is no tension crack zone as seen in the active case.

2) $c>0$

根据式(6-19),p_p随深度 z 的增加而线性增大。在被动土压力条件下,$z=0$ 处土压力为正,主动土压力条件下也不存在拉裂区。

The resultant E_p can be calculated as the summation of $E_{p,1}$(for a rectangular distribution) and $E_{p,2}$(for a triangular distribution), as seen in Figure 6-14.

E_p为$E_{p,1}$(矩形分布)和$E_{p,2}$(三角形分布)之和,如图 6-14 所示。

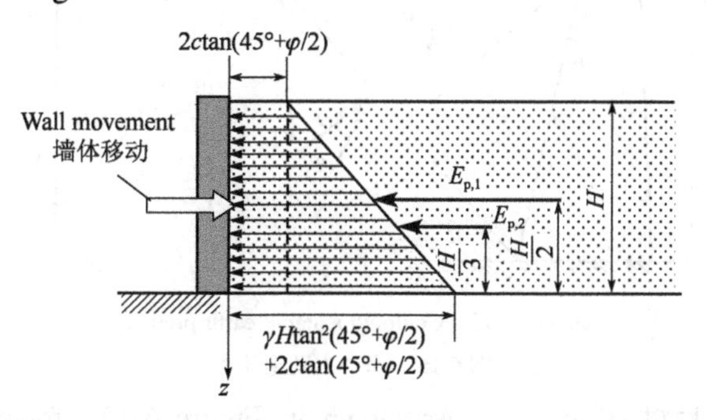

Figure 6-14 Rankine's passive earth pressure distribution ($c>0$)

图 6-14 朗肯被动土压力分布($c>0$)

$$E_{p,1} = 2c\tan\left(45° + \frac{\varphi}{2}\right) \times H \tag{6-19}$$

$$E_{p,2} = \frac{1}{2}\gamma H^2 \tan^2\left(45° + \frac{\varphi}{2}\right) \tag{6-20}$$

$E_{p,1}$ and $E_{p,2}$ are applied at heights of $H/2$ and $H/3$, respectively, from the base of the wall; thus, the center-of-gravity computation technique is used to determine the point of application of the total passive thrust $E_p(=E_{p,1}+E_{p,2})$.

$E_{p,1}$与$E_{p,2}$的作用点分别在距墙底 $H/2$ 和 $H/3$ 处,因此E_p($=E_{p,1}+E_{p,2}$)作用点位置在分布图形的形心处。

6.3 库仑土压力理论

库仑在1776年提出的土压力理论,其假定挡土墙后的填土为均匀砂土,当墙背离土体或推向土体时,墙后土体即达到极限平衡状态。库仑基于土楔的静力平衡条件,得到了作用在挡土墙上的主动土压力或被动土压力。

6.3.1 主动土压力

如图6-15所示,当刚性墙体向左移动时,墙后土体形成土楔*ABC*,*AB*和*BC*为破坏面。作用在土楔上的力包括土楔的重力*W*、土体作用在滑动面上的反力*R*和挡土墙对土楔的作用力E_a。考虑土楔的静力平衡条件,绘制*W*、*R*、E_a的力三角形。

6.3 Coulomb's earth pressure

Coulomb (1776) derived formulae to evaluate lateral earth pressures for sandy soils when a soil wedge behind a rigid wall slides due to a sufficient wall movement. He established the force equilibrium on the sliding soil wedge and derived the solution for the reaction force of the wall as an active or passive earth pressure.

6.3.1 Active case

When the rigid wall shown in Figure 6-15 moves to the left by a sufficient amount, a failing soil wedge *ABC* is formed, and lines *AB* and *BC* become failure surfaces. On the wedge, only three forces act, which are *W* (the weight of the wedge), *R* (the reaction force from the soil mass), and E_a (the reaction from the wall). These forces maintain an equilibrium condition, and the force polygon is closed, as seen in the right side of the figure.

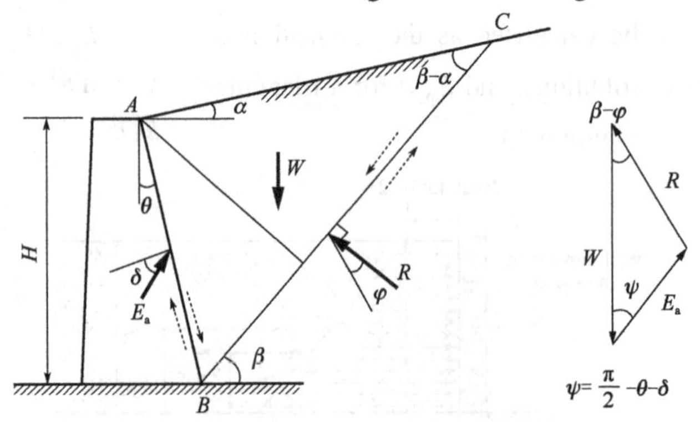

Figure 6-15 Coulomb's active earth pressure

图6-15 库仑主动土压力

在图6-15中,*R*与破坏面*BC*的法线成φ角,E_a与墙背法线成δ角,δ为墙背与填土间的摩擦角。土楔自重*W*垂直向下,三个力的方向和*W*值已知,因此E_a可由闭合的力三角形确定。E_a是墙背对土楔的反作用力,根据力的反作用原理,它等于土楔对墙背的主动土压力。

As shown in Figure 6-15, force *R* acts at an φ angle normal to the slide line *BC*, along which shear failure of the soil occurs. E_a acts at an angle (δ) inclined from the normal to the wall face, this angle is the friction angle between the sliding soil and the wall and is called the wall friction angle. Since *W* acts downward, the directions of all three forces and the magnitude of *W* are known; thus, the magnitude of E_a can be determined from a closed-force polygon. E_a is the reaction from the wall face against the sliding wedge. From the force-reaction principle, E_a is equal to the active earth thrust from the soil wedge against the wall at the active stage.

$$E_a = W\frac{\sin(\beta-\varphi)}{\sin(\beta-\varphi+\psi)} \tag{6-21}$$

In the construction shown in Figure 6-15, however, the failure angle β is unknown and has to be assumed. Thus, for an assumed angle β, the W and E_a value are obtained. Using several different β values, we obtain E_a and β relations as plotted in Figure 6-16. The maximum value among the trial E_a values is the active earth pressure according to the Coulomb method.

但根据图6-15和计算土压力时,破坏面位置是未知的,即β角是未知的。假定破坏面位置不同,相应的土楔重量W及土压力E_a均随之改变,如图6-16所示。当β等于某一值时,E_a达到最大值,称为库仑主动土压力。由此可知:

$$\frac{dE_a}{d\beta}=0$$

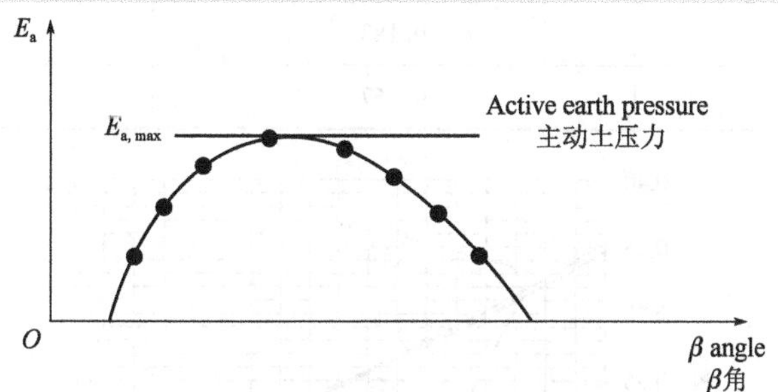

Figure 6-16 Active earth pressure determination by trial

图6-16 主动土压力的试算确定

Coulomb determined the analytical solution for the preceding active earth pressure E_a as:

库仑给出的主动土压力E_a的解析解为:

$$E_a = \frac{1}{2}\gamma H^2 \frac{\cos^2(\varphi-\theta)}{\cos^2\theta\cos(\delta+\theta)\left[1+\sqrt{\frac{\sin(\delta+\varphi)\sin(\varphi-\alpha)}{\cos(\delta+\theta)\cos(\theta-\alpha)}}\right]} = \frac{1}{2}\gamma K_a H^2 \tag{6-22}$$

$$K_a = \frac{\cos^2(\varphi-\theta)}{\cos^2\theta\cos(\delta+\theta)\left[1+\sqrt{\frac{\sin(\delta+\varphi)\sin(\varphi-\alpha)}{\cos(\delta+\theta)\cos(\theta-\alpha)}}\right]^2} \tag{6-23}$$

Where α and θ are defined in Figure 6-15.

式中,α和θ在图6-15中已定义。

Table 6-1 and Figure 6-17 show K_a values for a vertical wall ($\theta = 0$) with horizontal backfill ($\alpha = 0$) with $\delta = \varphi/2$ and $2\varphi/3$. From Figure 6-17, it can be seen that K_a decreases with increasing φ and that the effect of δ is small. Readers are encouraged to create their own spreadsheets to compute K_a values for other combinations of α, θ, φ, and δ values based on Equation (6-23).

表6-1和图6-17为填土顶面水平($\alpha = 0$)且$\delta = \varphi/2$和$2\varphi/3$竖直墙背($\theta = 0$)的K_a值。从图6-17可以看出,K_a随φ角的增大而减小,而墙面摩擦角δ的影响较小。建议读者根据式(6-23)创建电子表格,计算不同α、θ、φ和δ值对应的K_a值。

Table 6-1 Coulomb's K_a values for $\theta = 0$, $\alpha=0$, $\delta=\varphi/2$, and $2\varphi/3$

表 6-1 $\theta=0,\alpha=0,\delta=\varphi/2,2\varphi/3$ 时 K_a的值

φ	$K_a(\delta=\varphi/2)$	$K_a(\delta=2\varphi/3)$
26	0.353	0.347
28	0.326	0.321
30	0.301	0.297
32	0.278	0.275
34	0.256	0.254
36	0.236	0.235
38	0.217	0.217
40	0.199	0.200
42	0.183	0.184
44	0.167	0.167

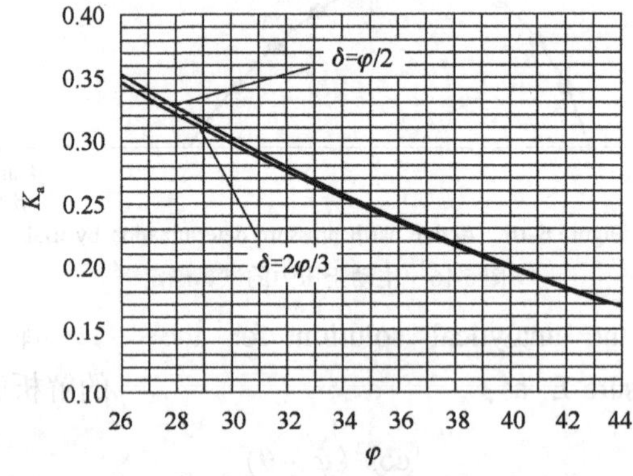

Figure 6-17 K_a for $\delta=\varphi/2$ and $2\varphi/3$ ($\alpha=0$ and $\theta=0$)

图 6-17 $\delta=\varphi/2$ 和 $2\varphi/3$ 的 K_a值($\alpha=0$ and $\theta=0$)

值得注意的是,根据式(6-23),当$\alpha=0$(填土顶面水平)、$\theta=0$(墙背竖直)和$\delta=0$(墙面光滑)时,库仑主动土压力系数K_a值与朗肯主动土压力系数相同[式(6-9)]。

It is noted that in Equation (6-23), when $\alpha=0$ (horizontal backfill), $\theta=0$ (a vertical wall), and $\delta=0$ (a smooth wall) are chosen, the K_a value obtained using Coulomb's solution becomes the same as that obtained by Rankine's formula [Equation (6-9)].

6.3.2 被动土压力

库仑被动土压力理论假设刚性墙体朝向墙背填土移动时会形成一个土楔,如图 6-18 所示。与图 6-15 中的主动土压力相比,在被动土压力条件下,土楔被挤向上滑动,反作用力 R 和 E_p偏向阻止土楔滑动的方向。

6.3.2 Passive case

Coulomb's passive earth pressure theory similarly assumes that a solid wedge is formed behind a rigid wall, which is moved against the soil mass until failure, as seen in Figure 6-18. Note that in the passive case, the wedge is pushed up so that the reactions R and E_p act from opposite directions relative to the faces of the sliding wedge, in contrast to the active case illustrated in Figure 6-15. Assuming the angle β, E_p is obtained from a closed-force polygon. Using several β values, the minimum value of E_p is determined and is considered the passive earth thrust.

假设破裂角度为β,根据闭合力多边形求得E_p。通过试算不同的β值,对应的E_p最小值视为被动土压力。

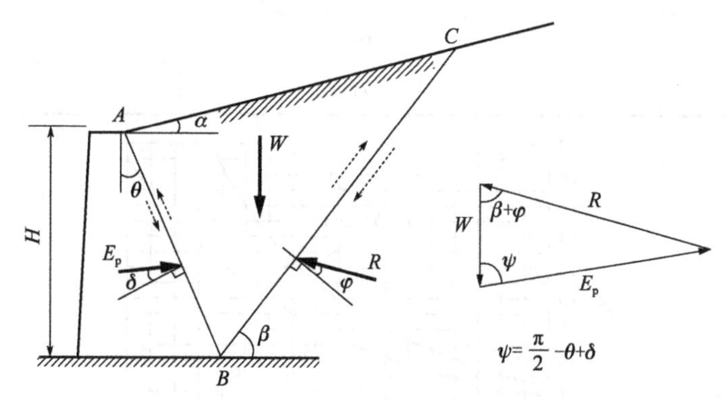

Figure 6-18 Coulomb's passive earth pressure

图6-18 库仑被动土压力

The analytical solution for Coulomb's passive earth pressure is given by:

库仑被动土压力的数值解为:

$$E_p = \frac{1}{2}\gamma H^2 \frac{\cos^2(\varphi+\theta)}{\cos^2\theta\cos(\delta-\theta)\left[1-\sqrt{\dfrac{\sin(\delta+\varphi)\sin(\varphi+\alpha)}{\cos(\delta-\theta)\cos(\theta-\alpha)}}\right]^2} = \frac{1}{2}\gamma K_p H^2 \tag{6-24}$$

$$K_p = \frac{\cos^2(\varphi+\theta)}{\cos^2\theta\cos(\delta-\theta)\left[1-\sqrt{\dfrac{\sin(\delta+\varphi)\sin(\varphi+\alpha)}{\cos(\delta-\theta)\cos(\theta-\alpha)}}\right]^2} \tag{6-25}$$

Table 6-2 and Figure 6-19 show K_p values for a vertical wall ($\theta = 0$) with horizontal backfill ($\alpha = 0$) for $\delta = \varphi/2$ and $2\varphi/3$. The K_p values obtained are much higher than the K_a values obtained for the active case, and some differences in the results between the $\delta = \varphi/2$ and $\delta = 2\varphi/3$ cases are observed.

表6-2和图6-19为填土顶面水平($\alpha=0$)、墙背竖直($\theta=0$),且$\delta=\varphi/2$和$2\varphi/3$时的K_p值。与K_a值相比,K_p值要高得多,并且$\delta=\varphi/2$和$2\varphi/3$时的K_p值存在一定差异。

Table 6-2 Coulomb's K_p values for $\theta=0$, $\alpha=0$, $\delta=\varphi/2$, and $2\varphi/3$

表6-2 $\theta=0,\alpha=0,\delta=\varphi/2,2\varphi/3$ 时K_p的值

φ	$K_p(\delta=\varphi/2)$	$K_p(\delta=2\varphi/3)$
26	3.787	4.400
28	4.325	5.154
30	4.976	6.108
32	5.775	7.337
34	6.767	8.957
36	8.022	11.154
38	9.639	14.233

Continue
续上表

φ	$K_p(\delta=\varphi/2)$	$K_p(\delta=2\varphi/3)$
40	11.771	18.737
42	14.662	25.696
44	18.714	37.270

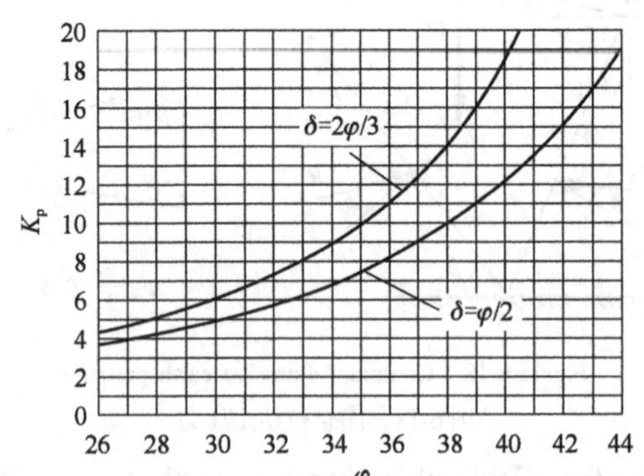

Figure 6-19 K_p for $\delta=\varphi/2$ and $2\varphi/3$ ($\alpha=0$ and $\theta=0$)

图 6-19 $\delta=\varphi/2$ 和 $2\varphi/3$ 的 K_p 值($\alpha=0$ and $\theta=0$)

值得注意的是,根据式(6-25)中,当 $\alpha=0$(填土顶面水平)、$\theta=0$(墙背竖直)和 $\delta=0$(墙面光滑)时,库仑被动土压力系数 K_p 值与朗肯被动土压力系数相同[式(6-17)]。

It is noted that for $\alpha=0$ (horizontal backfill), $\theta=0$ (a vertical wall), and $\delta=0$ (a smooth wall), Coulomb's K_p value from Equation (6-25) becomes the same as that given by Rankine's formula [Equation (6-17)].

6.3.3 墙后填土分层时土压力的计算

6.3.3 Calculation of earth pressure in layered fill behind wall

墙后填土分层时,因各层土具有不同的物理力学性质,须用近似方法分层计算土压力。

When the soil behind the wall is layered, an approximate method should be used to calculate the earth pressure in each layer because the different physical and mechanical properties of each layer of soil are different.

如图6-20所示,墙后填土水平时,计算时假设土的分层面与填土表面平行,先由库仑公式求得上一层的土压力 E_{a1},并画出其土压力强度分布图。

As shown in Figure 6-20, when the backfill behind the wall is a horizontal plane, the layered surface of the soil is assumed to be parallel to the surface of the fill. First, the earth pressure E_{a1} of the upper layer is obtained from Coulomb's formula, and the distribution diagram of the earth pressure intensity is drawn.

$$E_{a1}=\frac{1}{2}\gamma_1 H_1^2\cdot K_{a1} \tag{6-26}$$

式中,K_{a1} 为上层土的主动土压力系数。

Where K_{a1} is the active earth pressure coefficient of the upper layer of soil.

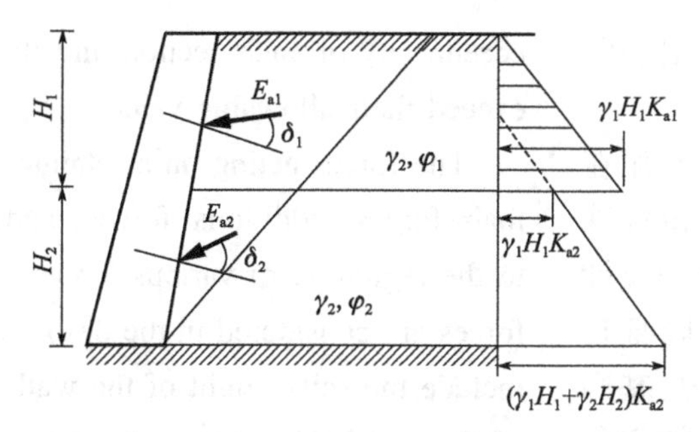

Figure 6-20 Coulomb's active earth pressure in layered soils

图 6-20 成层土中的库仑主动土压力

When calculating the earth pressure of the lower layer of soil, the weight of the upper soil is regarded as a uniformly distributed load. The earth pressure distribution graph is shown in Figure 6-21. The earth pressure of the lower layer of soil is E_{a2}, given by:

算下层土的土压力时，将上层土的重量视为均布荷载，按库仑公式作出土压力分布图，即为下层土的压力分布图，如图 6-21 所示。下层土的土压力为 E_{a2}。

$$E_{a2} = \left(\gamma_1 H_1 H_2 + \frac{1}{2}\gamma_2 H_2^2\right) K_{a2} \tag{6-27}$$

Where K_{a1} is the active earth pressure coefficient of the lower layer of soil.

式中，K_{a2} 为下层土的主动土压力系数。

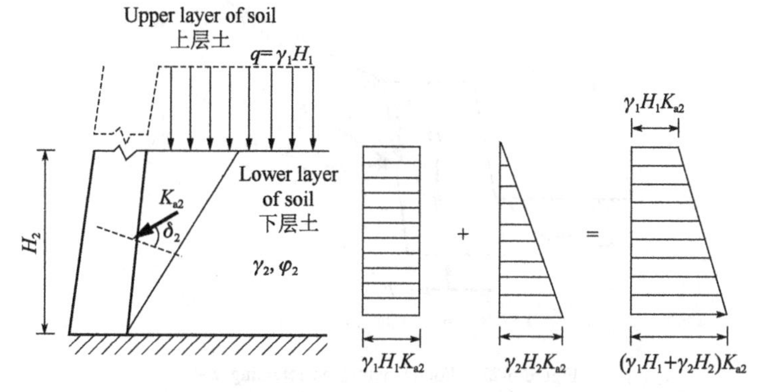

Figure 6-21 Coulomb's active earth pressure in layered soils

图 6-21 成层土中的库仑主动土压力

The positions of the action points of E_{a1} and E_{a2} can be determined from the centroid of the corresponding pressure graph, and the directions of the action points are shown, along with the respective δ_1 and δ_2 angles with the normal of the wall back.

E_{a1} 和 E_{a2} 的作用点位置可通过相应压力图形的形心来确定，方向则分别与墙背法线成 δ_1、δ_2 夹角。

6.4 Stability of gravity retaining wall

6.4 重力式挡土墙的稳定性

When a gravity retaining wall is used to bear earth pressure, it must have sufficient strength and bearing capacity to resist sliding or and overturning under the action of self-weight and external forces. In addition, the stress and

重力式挡土墙承受土压力时应具有足够的强度及稳定性，在自重和外力的作用下墙体不发生滑动和倾覆，同时，墙身每一截面和基底的应力及偏心距均不超过其容许值。

eccentricity of each section and the base of the wall cannot exceed their allowable values.

根据力的特性,可将作用在挡土墙的主力、附加力和特殊力。一般情况下,挡土墙的设计仅考虑主力,主力包括墙身自重、墙后土体主动土压力、基底反力和基底摩擦力等。特殊条件下(如洪水、地震等),还应考虑附加力和特殊力的作用。一般情况下,可忽略墙前土体的被动土压力。当基础埋置较深或对墙址前的地面进行防护时,为了确保墙前土体发挥作用,可部分或全部计入此抗力,一般按 $K_p=1$ 算被动土压力,如图 6-22 所示。

The forces acting on a retaining wall can be divided into main forces, additional forces, and special forces, according to the regularity of various forces. In general, only the main forces are considered in the design of a retaining wall. These include the self-weight of the wall, the active earth pressure of the soil behind the wall, the reaction force of the base, and the friction force of the foundation. Under special conditions (such as a flood or an earthquake), additional forces and special forces should be considered. In addition, the passive earth pressure on the soil in front of the wall is often ignored. However, when the foundation is buried deep or when the ground in front of the wall is protected, this resistance can be partially or fully considered. In general, the passive earth pressure is calculated assuming $K_p = 1$, as shown in Figure 6-22.

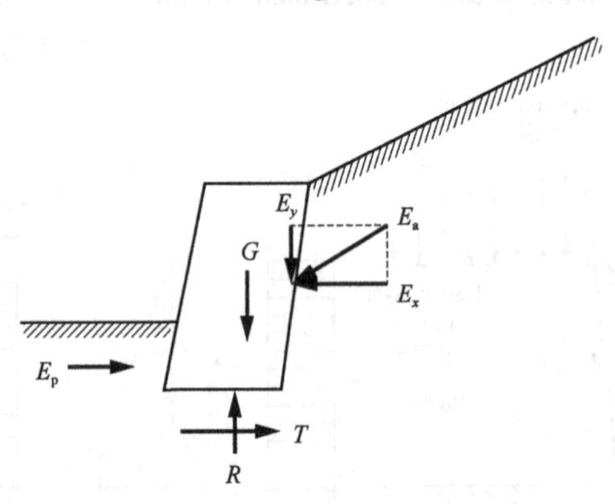

Figure 6-22　Forces acting on retaining wall

图 6-22　挡土墙的作用力系

对于重力式挡土墙,墙的整体稳定性往往是挡土墙设计中的控制因素。挡土墙的稳定包括抗滑稳定及抗倾覆稳定。

For a gravity retaining wall, the overall stability is typically the controlling factor in the design. The overall stability of a retaining wall depends on both its anti-sliding stability and anti-overturning stability.

1)抗滑稳定验算

1) Anti-sliding stability verification

挡土墙的抗滑稳定性是指在土压力及其他外力作用下,基底摩擦力抵抗挡墙滑移的能力。即作用于挡墙的最大抗滑力与实际滑动力之比,以抗滑稳定系数 K_c 表示,如图 6-23 所示。

The anti-sliding stability of a gravity retaining wall refers to the capacity of the base of the wall to resist sliding under the action of earth pressure and other external forces. The ratio of the maximum possible anti-sliding force acting on the retaining wall to the actual sliding force gives the anti-sliding stability coefficient K_c, as shown in Figure 6-23.

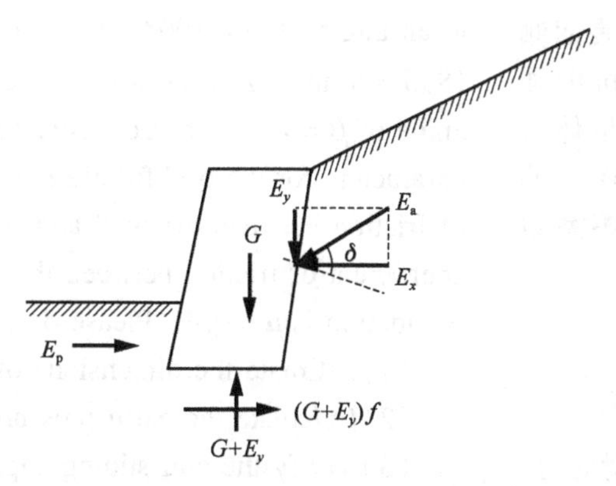

Figure 6-23 Forces acting on retaining wall for anti-sliding stability verification

图 6-23 挡土墙抗滑稳定性检算受力示意图

$$K_c = \frac{(G+E_y)\cdot f + E_P}{E_x} \tag{6-28}$$

Where E_x and E_y are the horizontal and vertical components of the active earth pressure, respectively, at the back of the wall; f is the friction coefficient of the foundation, which can be determined from field testing or estimated using empirical data; and E_p is the passive earth pressure, $E_p = 0$ when the passive earth pressure is not considered.

式中,E_x、E_y为墙背主动土压力的水平及竖直分力;f为基底摩擦系数,可通过现场试验或参照经验数据确定;E_p为被动土压力,如不计被动土压力时,$E_p = 0$。

To ensure the anti-sliding stability of a retaining wall, $K_c \geqslant 1.3$ is specified. K_c should not be less than 1.2 when additional force is considered.

为保证挡土墙的抗滑稳定性,规定 $K_c \geqslant 1.3$。考虑附加力时,K_c不应小于1.2。

2) Anti-overturning stability verification

2)抗倾覆稳定验算

The anti-overturning stability of a retaining wall refers to the capacity of the wall body to resist overturning around the toe of the wall. The ratio of the sum of the stability moments to the sum of the overturning moments, expressed by the anti-overturning stability coefficient K_0, is shown in Figure 6-24. The anti-overturning stability coefficient K_0 is calculated as:

挡土墙的抗倾覆稳定性是指它抵抗墙身绕墙趾向外转动倾覆的能力。即整个挡土墙绕墙趾的稳定力矩之和与倾覆力矩之和的比值,用抗倾覆稳定系数 K_0表示,如图 6-24 所示。抗倾覆稳定系数 K_0的计算方法为:

$$K_0 = \frac{G\cdot L_G + E_y\cdot L_y + E_P\cdot L_P}{E_x\cdot L_x} \tag{6-29}$$

To ensure the anti-overturning stability of a retaining wall, K_0 must be $\geqslant 1.5$, and it should not be less than 1.3 when additional force is considered.

为保证挡土墙的抗倾覆稳定性,须满足 $K_0 \geqslant 1.5$,考虑附加力时,K_0不应小于1.3。

Example 6-3

例题 6-3

Details of a retaining wall are shown in Figure 6-25. The water table is below the base of the wall. The height of the wall is $H = 6.0$ m, and the back of the wall is inclined

某挡土墙如图 6-25 所示,地下水位低于墙底。墙高 $H = 6.0$ m,墙背倾斜 $\alpha = 100°$。回填

土重度 $\gamma = 18.5\ kN/m^3$,表面倾斜角度 $\beta = 10°$。回填土抗剪强度参数 $c = 0$ 和 $\varphi = 30°$。墙体与回填土之间的摩擦角为 20°。墙体底部与地基之间的摩擦系数为 $\mu = 0.4$。请设计墙壁。

(1)估算墙体尺寸。

(2)计算土压力。

(3)验算挡土墙的抗滑能力。

(4)验算挡土墙的抗倾覆能力。

at an angle of $\alpha = 100°$. The unit weight of the backfill is $18.5\ kN/m^3$. The surface of the backfill is inclined at an angle of $\beta = 10°$. Characteristic values of the shear strength parameters for the backfill are $c = 0$ and $\varphi = 30°$. The angle of friction between the wall and the backfill is $\delta = 20°$. The coefficient of friction between the bottom of the wall and the foundation is $\mu = 0.4$. Please design a retaining wall.

(1) Estimate the dimensions of the wall.

(2) Calculate the earth pressure.

(3) Verify the anti-sliding capacity of the wall.

(4) Verify the anti-overturning capacity of the wall.

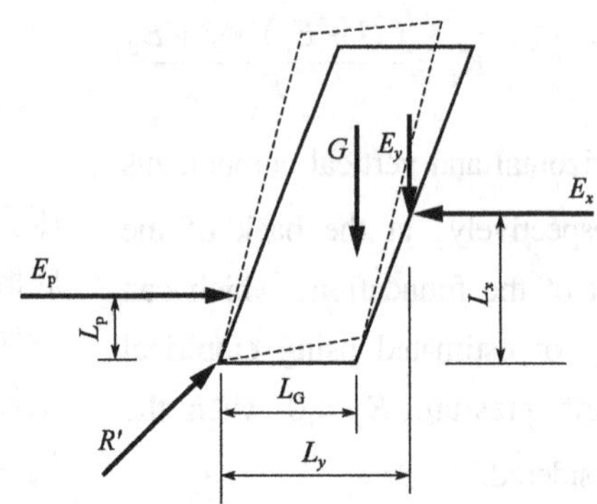

Figure 6-24 Forces acting on retaining wall for anti-overturning stability verification

图 6-24 挡土墙抗倾覆稳定性检算受力示意图

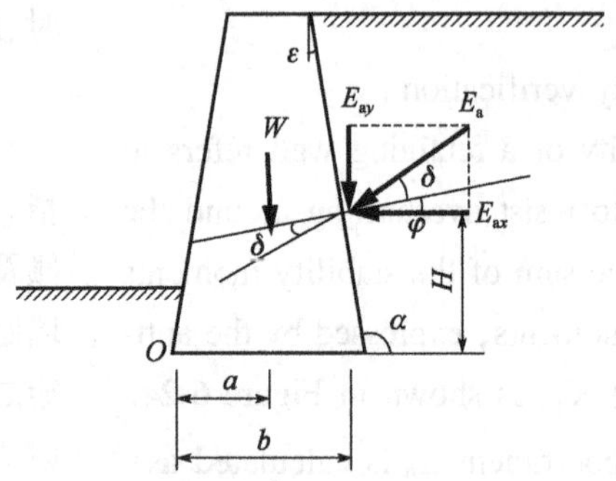

Figure 6-25 Diagram for Example 6-3

图 6-25 例题 6-3 的附图

解:

(1)设挡土墙的上下表面宽度分别为 1.0 m 和 5.0 m。因此,墙体自重为:

Solution

(1) Let the width of the top and bottom surface be 1.0 m and 5.0 m, respectively. Thus, the self-weight of the wall is:

$$W = \frac{(1.0 + 5.0) H\gamma}{2} = 333(\text{kN/m})$$

(2)对于 $\alpha = 100°$, $\varepsilon = 10°$, $\varphi = 30°$, $\delta = 20°$,由式 6-23 得到 $K_a = 0.377$。

(2) For $\alpha = 100°$, $\varepsilon = 10°$, $\varphi = 30°$, and $\delta = 20°$, from Equation (6-23), we obtain $K_a = 0.377$.

According to Equation (6-27):

根据式 6-27 有:

$$E_a = \frac{1}{2}K_a\gamma H^2 = 125.5(\text{kN/m})$$

The vertical component of the earth pressure is:

土压力的垂直分量为:

$$E_{ay} = E_a\sin(\varepsilon+\delta) = 62.8(\text{kN/m})$$

The horizontal component of the earth pressure is:

土压力的水平分量为:

$$E_{ax} = E_a\cos(\varepsilon+\delta) = 108.7(\text{kN/m})$$

(3) Anti-sliding capacity verification

(3)抗滑稳定验算

$$K_c = \frac{(W+E_{ay})\mu}{E_{ax}} = 1.46 > 1.3$$

This solution is safe, and the soil has anti-sliding stability, but the factor of safety is extremely large. Let the width of the bottom surface be 4.5 m. The self-weight of the wall is then $W' = (1.0\text{m} + 4.5\text{m})H\gamma/2 = 305.25$ kN/m. As a result:

这种方法计算的结果是安全的。且土体具有抗滑稳定性,且安全系数极大。底面宽度设为 4.5 m。此时墙体自重为 $W' = (1.0\text{ m} + 4.5\text{ m})H\gamma/2 = 305.25$ kN/m。则:

$$K_c = \frac{(W'+E_{ay})\mu}{E_{ax}} = 1.35 > 1.3$$

(4) Anti-overturning capacity verification

(4)抗倾覆稳定验算

The lever arms for the forces against the toe of the wall O are:

对墙趾的力臂为:

Lever arm for W': $a = 2.25$ m;

Lever arm for E_{ay}: $b = 5.08$ m;

Lever arm for E_{ax}: $h = 2.00$ m.

力臂为 W': $a = 2.25$ m;

力臂 E_{ay}: $b = 5.08$ m;

力臂 E_{ax}: $h = 2.00$ m。

Using Equation (6-29):

结合式(6-29),有:

$$K_0 = \frac{W'a + E_{ay}b}{E_{ax}h} = 4.6 \geqslant 1.5$$

Thus, the soil meets the requirements for anti-overturning capacity.

因此,土体满足抗倾覆稳定要求。

Exercises

习 题

(1) The back of a retaining wall is smooth and vertical; the fill surface is horizontal; the wall height is 6 m; the fill is a homogeneous cohesive soil with an internal friction angle of 30°, a cohesion $c = 8.67$ kPa, and a unit weight of $\gamma = 20$ kN/m^3. Calculate the resultant force of the active earth pressure acting on the wall back.

(1)某墙背光滑、直立,填土面水平,墙高 6 m。填土为均质黏土,其内摩擦角为 30°、黏聚力为 8.67 kPa、重度为 $\gamma = 20$ kN/m^3,计算作用在墙背上的主动土压力合力。

(2)某墙背直立、光滑,填土面水平的挡土墙,高4 m。填土为均质黏土,其内摩擦角 $\varphi=20°$、黏聚力 $c=10$ kPa、重度 $\gamma=17$ kN/m^3,侧向压力系数 $K_0=0.66$。若挡土墙没有位移,计算作用在墙上土压力合力 E_0 大小及其作用点距墙底的位置 h。

(2) A retaining wall with a smooth and vertical back and horizontal fill surface is 4-m-high. The fill is a homogeneous cohesive soil with an internal friction angle 20°, cohesion $c=10$ kPa, unit weight $\gamma=17$ kN/m^3, and lateral pressure coefficient $K_0=0.66$. Assuming that there is no displacement of the retaining wall, calculate the resultant force E_0 of earth pressure acting on the wall and the position h from the point of action to the bottom of the wall.

(3)某挡土墙高5 m,假定墙背垂直光滑,墙后填土面水平。填土的黏聚力 $c=11$ kPa,内摩擦角 $\varphi=20°$,重度 $\gamma=18$ kN/m^3,试计算墙背主动土压力分布图形和主动土压力的合力。

(3) The height of a retaining wall is 5 m. Assume that the back of the wall is smooth and vertical; the fill surface behind the wall is horizontal; the cohesion of the fill soil is $c=11$ kPa; the internal friction angle is $\varphi=20°$; and the unit weight is $\gamma=18$ kN/m^3. Calculate the distribution pattern of the active earth pressure on the back of the wall and the resultant active earth pressure.

(4)某挡土墙高5 m,墙背垂直光滑,填土面水平。$\gamma=18.0$ kN/m^3,$\varphi=22°$,$c=15$ kPa,试计算:①该挡土墙主动土压力分布、合力大小及其作用点位置;②若该挡土墙在外力作用下,朝填土方向产生较大的位移时,作用在墙背的土压力分布、合力大小及其作用点位置又为多少?

(4) The height of a retaining wall is 5 m; the back of the wall is smooth and vertical; the fill surface is horizontal; $\gamma=18.0$ kN/m^3, $\varphi=22°$, and $c=15$ kPa. ① Calculate the active earth pressure distribution, resultant force, and its action point position on the retaining wall. ② If the retaining wall exhibits a large displacement in the fill direction under the action of an external force, what is the earth pressure distribution, resultant force, and its action point position on the back of the wall?

(5)某挡土墙高度为10 m,墙背竖直,墙后填土水平。填土上作用均布荷载 $q=20$ kPa,墙后填土分两层:上层为中砂,重度 $\gamma_1=18.5$ kN/m^3,内摩擦角 $\varphi_1=30°$,层厚 $h_1=3$ m;下层为粗砂,$\gamma_2=19$ kN/m^3,$\varphi_2=35°$,地下水位在离墙顶6 m处,水下粗砂的饱和重度为 $\gamma_{sat}=20$ kN/m^3。计算作用在此挡土墙上的总主动土压力和水压力,并绘制压力分布图。

(5) The height of a retaining wall is 10 m; the back of the wall is vertical; the fill behind the wall is horizontal, and the uniformly distributed pressure on the fill is $q=20$ kPa. The fill behind the wall is divided into two layers: an upper layer of medium sand ($\gamma_1=18.5$ kN/m^3, internal friction angle $\varphi_1=30°$, $h_1=3$ m) and a lower layer of coarse sand ($\gamma_2=19$ kN/m^3, $\varphi_2=35°$). The groundwater level is located 6 m away from the top of the wall, and the saturated weight of the underwater coarse sand is $\gamma_{sat}=20$ kN/m^3. Calculate the total active earth pressure and water pressure acting on the retaining wall, and draw the pressure distribution diagram.

Slope Stability 边坡稳定性 7 Chapter

7.1 Introduction

Slopes in soils and rocks are ubiquitous in natural and man-made structures. Railways, highways, dams, levees, and canals are constructed by sloping the lateral faces of the soil because building slopes is generally less expensive than constructing walls.

Natural forces (wind, water, snow, etc.) change the topography of land, often creating unstable slopes. Failures of natural slopes (landslides) and man-made slopes usually results in considerable loss of life, destruction of property, economic losses, and environmental damage. These failures can be sudden and catastrophic or insidious and widespread or localized.

The collapse of a slope can be modeled by a block on a slope, as shown in Figure 7-1. The block is a failed soil mass, and the surface of the slope is the sliding surface. In the model, the force component (T) of the block in the direction of the slope activates the sliding. The reaction force to the weight of the block W is F on the sliding surface. The frictional component (F_x) of F in the direction of the slope is the force that resists the sliding. Until sliding occurs, T and F_x are balanced. The limiting value of F_x depends on the soil's shear strength τ_f. When τ_f is fully activated on the

7.1 概述

在天然的或人工的结构中土质边坡和岩质边坡是常见的。因土质边坡造价低于结构墙,因此土质边坡是较为常用的一种结构形式,如铁路、公路、水坝、堤坝和运河等建筑中多采用边坡形式。

自然营力(风、水、雪等)会改变地表形态,常常形成不稳定的边坡。天然边坡(山体滑坡)和人工边坡的失稳通常会导致人员伤亡、财产破坏、经济损失和环境破坏。有些边坡失稳是骤然的和灾难性的,有些则是隐伏的。有些边坡失稳规模较大,有些则是局部的。

边坡的崩塌可以通过斜坡上的一个块体模拟,如图 7-1 所示。上部块体表示已破坏的土体,斜坡顶面表示滑动面。在模型中,块体在斜坡方向上的分力(T)为滑动力。滑动面上对块体重量 W 的反作用力为 F。F 在斜坡方向上的摩擦分力(F_x)为抵抗块体滑动的力。在块体滑动之前,T 和 F_x 保持平衡,这样就不会发生滑动。F_x的极限值与土体的抗剪强度τ_f有关,当τ_f在滑动面上完全发挥时,上部块体就会发生滑动。

sliding surface, a slide will occur.

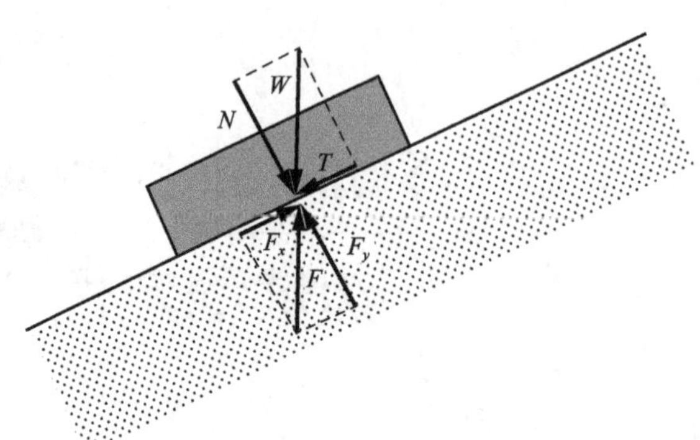

Figure 7-1 Block model for slope failure

图 7-1 边坡失稳的块体模型

大量观察资料表明,黏性土边坡破坏时的滑动面近似圆柱,横断面呈圆弧形;砂土边坡的破坏面近似平面,横断面呈直线,如图 7-2 所示。抗滑安全系数为:

Substantial observational data show that the sliding surfaces of cohesive soil slopes are similar to cylindrical surfaces and have a circular cross section. The fracture surface of a sandy soil slope is approximately a plane and has a straight-line cross section, as shown in Figure 7-2. The factor of safety (F_s) against sliding is given by:

$$F_s = \frac{\sum \tau_f}{\sum \tau} \tag{7-1}$$

$$F_s = \frac{\sum \tau_f \cdot r}{\sum \tau \cdot r} \tag{7-2}$$

式中,τ_f为沿滑动面土体的抗剪强度;τ为滑动面上的剪应力;r为圆形滑动面的半径;式(7-1)中的F_s表示为滑动面上抗剪强度与剪切应力的比值;式(7-2)中的F_s是抗滑力矩与滑动力矩的比值。

In these equations, τ_f is the shear strength of soil along the sliding surface; τ is the mobilized shear stress on the sliding surface, and r is the radius of the circular sliding surface. The F_s in Equation (7-1) is expressed as the ratio of the shear strength to the mobilized shear stress on the sliding surface. The F_s in Equation (7-2) is expressed as the ratio of the resisting moment to the driving moment.

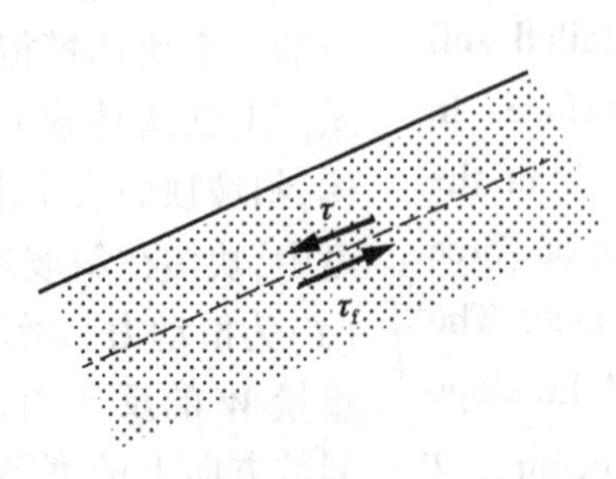

a) Plane sliding surface

a)平面滑动面

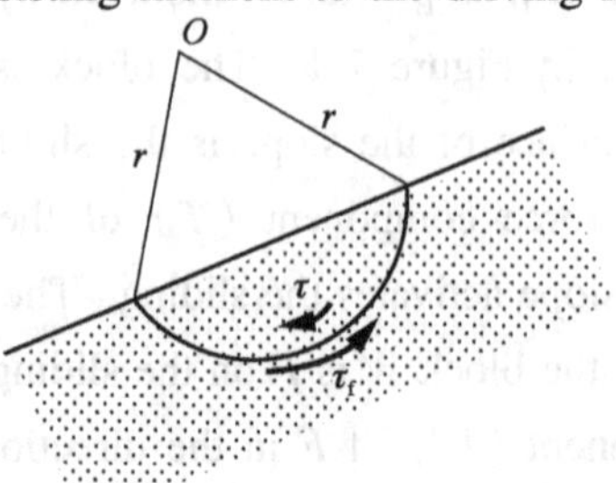

b) Circular sliding surface

b)圆弧滑动面

Figure 7-2 Definitions of factor of safety against slope failure

图 7-2 边坡失稳安全系数的定义

In railway design and safety evaluation, to consider the influence of a train load and permanent loads, such as those of the rails, sleepers, and ballast, on the subgrade surface, the loads are usually converted into a soil column with the same weight and the same type of soil as the subgrade, this soil column is called a conversion soil column, which acts on the top surface of subgrade. When checking the stability of a soil slope, it is necessary to add the conversion soil column on the top of subgrade as a part of the soil slope and then perform calculations according to the conventional evaluation method.

在铁路设计和安全评估中,为了研究列车荷载和永久荷载(如钢轨、枕木和道砟的荷载)对路基的影响,通常将路基表面的荷载转换成与路基重量和土体相同的土柱作用在路基顶面,称之为换算土柱。在验算土坡稳定性时,需要将换算土柱添加到路基顶面,作为土坡的一部分,然后根据常规评估方法进行计算。

An embankment is typically a strip structure, with a longitudinal length much larger than the cross-sectional width. Therefore, the stability of an embankment soil slope can be checked by treating it as a plane strain problem, i. e. , a unit longitudinal length can be considered in the calculation.

路堤一般都是条形结构,其纵向长度远远大于横截面的宽度。因此,可以将路堤土坡的稳定性检查视为平面应变问题,即在计算中可以考虑单位纵向长度。

7.2 Linear slip surface method

7.2 直线滑动面法

7.2.1 Cohesionless soil slopes with no seepage action

7.2.1 无渗流作用的无黏性土土坡

For embankments filled with permeable soil, such as sand, gravel, and crushed stone, which have no cohesive force c and have large internal friction angles φ, the linear slip surface method can be used to check the stability of a soil slope. The linear slip surface method assumes that the slip surface is a plane along which the unstable soil will slide as a whole.

对于用透水土填筑的路堤,如砂、砾石和碎石等,这类填料没有黏聚力 c,仅具有较大内摩擦角 φ,在进行土坡稳定性验算时,可采用直线滑动面法。直线滑动面法假定滑动面为一平面,破坏时不稳定的土体将沿此平面作整体滑动。

As shown in Figure 7-3, it is assumed that the unstable soil mass ABC slides along the plane AB and that the sum of the self-weight of the soil mass ABC and the external load is$W$. If the horizontal inclination angle of AB is α, then the force component T of weight W along AB, which is called the sliding force, pushes the soil downward. The shear strengths on AB, i. e. , the cohesion c and friction resistance on AB, are called the anti-sliding forces.

如图 7-3 所示,假定不稳定土体 ABC 沿 AB 平面滑动,ABC 土体自重与外荷载之和为 W。AB 平面与水平面之间的倾角为 α,则 W 沿 AB 面的分力 T 推动土体下滑,称为下滑力。AB 面上的的抗剪强度,即黏聚力 c 及摩阻力,统称为抗滑力。

The sliding force is given by:

下滑力为:

$$T = W \cdot \sin\alpha \tag{7-3}$$

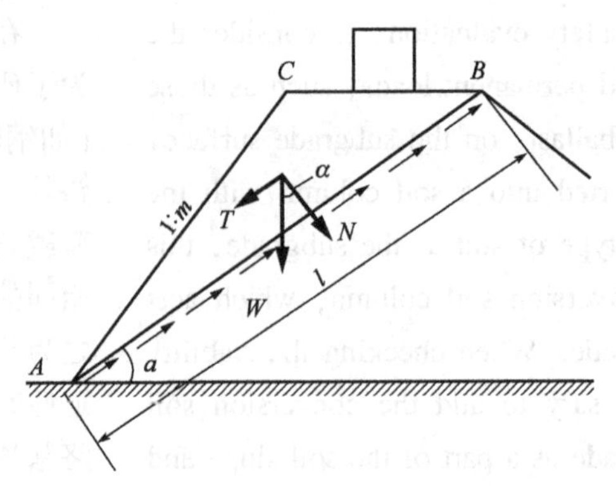

Figure 7-3 Slope stability analysis of cohesionless soil

图 7-3 无黏性土的土坡稳定性分析

抗滑力为：

The anti-sliding force is given by:

$$D = W \cdot \cos\alpha \cdot \tan\varphi + c \cdot l \tag{7-4}$$

如前所述,土坡稳定性系数是以滑动面上抗滑力与滑动面上滑动力的比值来定义的,因此:

As mentioned previously, the stability coefficient of a soil slope is defined as the ratio of the anti-sliding force to the sliding force at the failure surface. Thus:

$$FS = \frac{W \cdot \cos\alpha \cdot \tan\varphi + c \cdot l}{W \cdot \sin\alpha} \tag{7-5}$$

式中,W 为滑体 ABC 的自重(kN);α 为滑动面的倾角(°);l 为滑动面 AB 长度(m)。

Where W is the weight of the sliding mass ABC (kN); α is the horizontal inclination angle of the failure surface (°), and l is the length of the failure surface AB (m).

因此,当上式中 $c=0$(粗粒土)时,抗滑安全条件为 $F_s > 1.0$,因此:

Accordingly, when $c = 0$ (coarse-grained soils) in the preceding equation, the condition for safe against sliding is $F_s > 1.0$. Thus,

$$\tan\alpha < \tan\varphi, \alpha < \varphi \tag{7-6}$$

当 $c=0$ 时,若坡角 α 小于土的内摩擦角 φ,则认为边坡是安全的。这一规律可由以下现象证明:当干砂从空中落下时,会形成一个倾角为 φ 的砂锥,φ 是砂粒在最松散状态下的最小内摩擦角,称为休止角。

In the case of $c = 0$, if the slope angle α is less than the soil's internal friction angle φ, the slope is considered safe. This is demonstrated by the fact that when dry sand is dropped gently from the air, it forms a sand cone with a side angle of φ, which is the smallest angle of internal friction for the loosest deposit and is called the angle of repose.

对于坡度较小的路堤,可以按照以下步骤进行边坡稳定性计算。在路堤边坡上取 A_1、A_2……若干点,如图 7-4 所示。假设 BA_1 和 BA_2 等许多可能的破坏面,并求出相应的 K 值。根据 α、K 值,画出 K-α 线,作一平行

For an embankment with a small transverse slope, the slope stability calculation can be carried out according to the following steps. Take points A_1 and A_2 on the embankment slope. Many failure surfaces can be assumed through BA_1 and BA_2, as shown in Figure 7-4, and the corresponding K values can be calculated. Based on the respective α and K values, draw a K-α curve, and make a straight line parallel to the α axis and tangent to the K-α curve. The α and K values associated with the tangent point represent the horizontal inclination angle and stability coefficient value of the most dangerous failure surface. When the transverse slope of the foundation is large, the possibility of the whole embankment sliding along the base surface should also be checked.

α 轴的直线，与 K-α 曲线相切时的 α、K 值表示最危险滑动面的水平倾角与稳定系数值。当路堤坡度较大时，还应验算整个路堤沿路堤底面滑动的可能性。

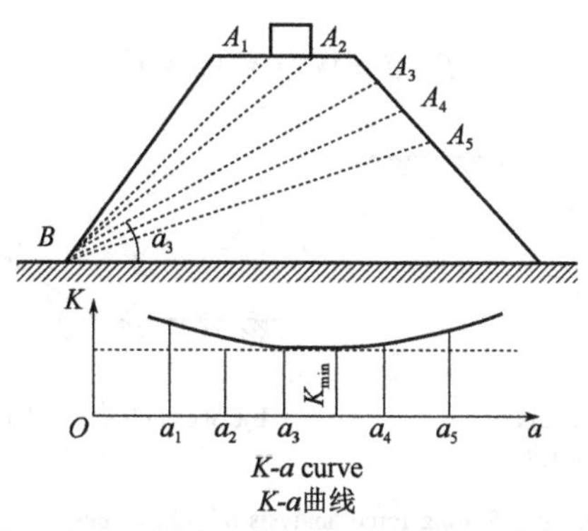

Figure 7-4 Determination of the most dangerous failure surface

图 7-4 最危险滑动面的确定

Example 7-1

A long natural slope in over-consolidated fissured clay with a saturated unit weight 20 kN/m^3 is inclined at 12° to the horizontal. A slip has developed on a plane parallel to the surface at a depth of 5 m. Determine the factor of safety along the slip plane using ①the critical-state parameter ϕ_{tu} = 28° and ②the residual strength parameter φ'_r = 20°.

例题 7-1

某天然土坡的饱和重度为 20 kN/m^3，其与水平面夹角呈 12°。在 5 m 深度处与表面平行的平面上形成滑移。分别采用①临界状态参数 ϕ_{tu} = 28°和②残余强度参数 φ'_r = 20°来计算沿滑动面的安全系数。

Solution

Equation(7-5) applies in both cases.

解：

式（7-5）适用于这两种情况。

①In terms of the critical-state strength:

①临界状态强度：

$$F_s = \frac{10.2\tan 28°}{20\tan 12°} = 1.28$$

②In terms of the residual strength:

②残余强度：

$$F_s = \frac{10.2\tan 20°}{20\tan 12°} = 0.87$$

7.2.2 Cohesionless soil slope with seepage

7.2.2 有渗流作用的无黏性土土坡

When reservoir impoundments or reservoir water levels drop suddenly, dams are affected by seepage forces, which have adverse effects on dam stability. A unit body is taken

当水库蓄水或水位快速下降时，坝体会受到渗透力的影响，这对坝体的稳定性产生不利影响。此时以土骨架为隔离体,在坡面上渗流逸出点以下取单元体。它除了本身重量外,还受到渗流力 J 的作用,如图 7-5 所示。因渗流方向与坡面平行,渗透力与坡面平行,渗流力的方向也与坡面平行,此时土体下滑力为:

below the seepage escape point on the slope. In addition to its self-weight, it is also affected by the seepage force J, as shown in Figure 7-5. The seepage direction is parallel to the slope surface; hence, the seepage force is parallel to the slope surface, and the direction of the seepage force is also parallel to the slope surface. The sliding force is given by:

$$T + J = W' \cdot \sin\alpha + J \tag{7-7}$$

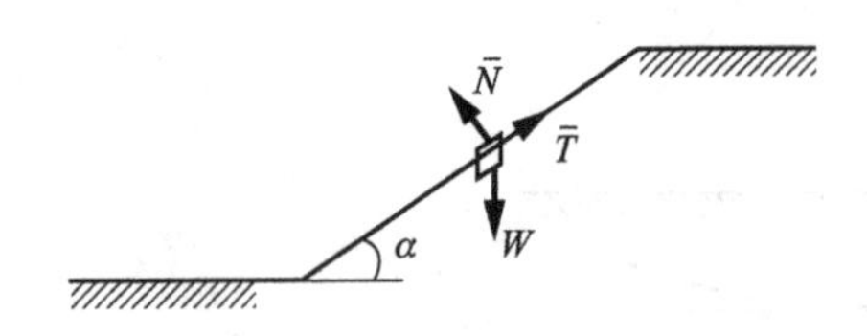

a)General cohesionless soil
a)一般无黏性土坡

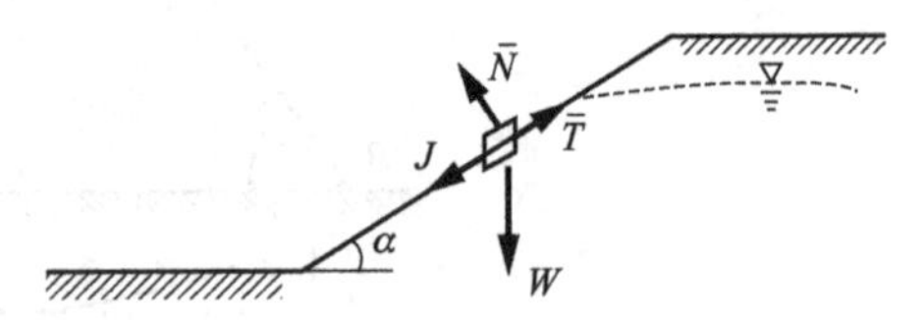

b)Cohesionless soil slope with flow along the slope
b)有顺坡渗流无黏性土坡

Figure 7-5 Sliding force analysis of cohesionless soil slope
图 7-5 无黏性土坡的平面滑动受力分析

此时,单元体所能发挥的最大抗滑力仍为 $D = W' \cdot \cos\alpha \cdot \tan\varphi$,于是安全系数为:

In this case, the maximum anti-sliding force of the unit is still $D = W' \cdot \cos\alpha \cdot \tan\varphi$; thus, the safety factor is:

$$F_s = \frac{D}{T + J} = \frac{W'\cos\alpha \cdot \tan\varphi}{W' \cdot \sin\alpha + J} \tag{7-8}$$

对于单元土体来说,如果直接考虑渗透效应,单元体积的土体自重就是浮重度 γ',而单位体积的渗流力 $j = i\gamma_w$,其中 γ_w 为水的重度;i 是水力梯度,$i = \sin\alpha$。因为是顺坡出流,上式可以写为:

For the unit soil, if the seepage force is used to consider the seepage effect directly, the unit weight of the soil mass is the buoyant weight γ', and the seepage force is $j = i\gamma_w$, where γ_w is the unit weight of water and i is the hydraulic gradient, $i = \sin\alpha$. Because water flows along the slope, the above formula can be written as:

$$F_s = \frac{\gamma'\cos\alpha \cdot \tan\varphi}{(\gamma' + \gamma_w)\sin\alpha} = \frac{\gamma'\tan\varphi}{\gamma_{sat}\tan\alpha} \tag{7-9}$$

7.3 条分法

7.3 Slope stability of slice method

7.3.1 计算条分法中 F_s

当边坡几何形状复杂且土质不均匀时,通常采用条分法计算安全系数。如图 7-6a)所示,假设一个以 O 点为圆心的圆弧滑动面,用垂直分界线将滑体分为若干条块。以第 4 个条块为

7.3.1 Calculation of F_s value in slice method

The slice method is often used when the geometry of slopes is complex and the soil's properties are not homogeneous. In Figure 7-6a), a circular slip surface with Point O as the center of rotation is assumed, and the slip mass is divided into several masses by vertical dividing lines. As an example, the 4th slice is picked to show all forces acting on it in Figure 7-6b). W is the total weight of the slice; E_i and E_{i+1} are normal forces on the vertical faces of the slice; T_i and T_{i+1} are the shear forces acting on the vertical faces; S_i is the shear resistance on the sliding surface; R'_i is the effective normal stress on the sliding surface, and U_i is the water pressure acting on the sliding surface. The shear strength along the sliding surface S_i is given by $S_i = c_i \cdot a_i + R'_i \cdot \tan\varphi_i$, where a_i is the sliding surface length for the ith slice. Among all these forces, i. e., W, E_i, E_{i+1}, T_i, T_{i+1}, R'_i, S_i, and U_i, the values of W, S_i, and U_i can be computed. The values of the remaining five forces, E_i, E_{i+1}, T_i, T_{i+1}, and R'_i, are unknown. Since only three equilibrium equations (i. e., $\sum H=0$, $\sum V=0$, and $\sum M=0$) are available for use in the solution, this problem is indeterminate.

例,受力如图7-6b)所示,W为条块的总重量,E_i和E_{i+1}为作用于条块垂直面上的法向力,T_i和T_{i+1}为作用于垂直面上的剪切力,S_i为滑动面上的剪切力,R'_i为滑动面上的有效法向应力,U_i为作用于滑动面上的水压。滑动面抗剪强度S_i的计算方法为$S_i = c_i \cdot a_i + R'_i \cdot \tan\varphi_i$,其中$a_i$是第$i$个条块的滑动面长度。在上述力中,$W$、$E_i$、$E_{i+1}$、$T_i$、$T_{i+1}$、$R'_i$、$S_i$和$U_i$,$W$、$S_i$和$U_i$的值可以计算得到,其余$E_i$,$E_{i+1}$,$T_i$,$T_{i+1}$和$R'_i$为5个未知力。由于只有3个平衡方程(即$\sum H=0$、$\sum V=0$和$\sum M=0$)可以求解,因此无固定解。

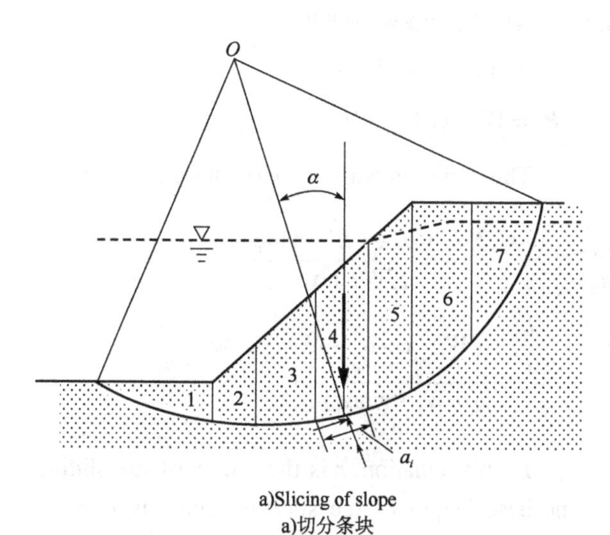

a)Slicing of slope
a)切分条块

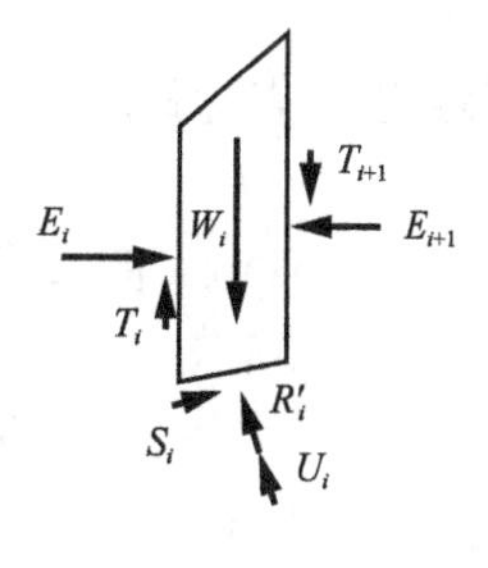

b)Forces acting on the 4th slice
b)力作用于第4个条块

Figure 7-6 Principle of slice method

图7-6 条分法计算原理

To solve indeterminate problems, some conditions must be assumed. In this text, the simplest technique, the ordinary slice method (swedish slice method), is introduced.

The ordinary slice method (Fellenius 1936), illustrated in Figure 7-7a), assumes that the resultants of T and E on two facing vertical faces of the slice are equal and act in opposite directions on the same line. Thus, T and E are eliminated from the force and moment equilibrium computation. All other forces pass through the midpoint on the sliding

为了解决上述不确定问题,需假定一些条件。本书介绍最简单的瑞典条分法。

如图7-7a)所示,瑞典条分法(Fellenius 1936)假设条块两个垂直面上的T和E大小相等、方向相反,因此T和E在力和力矩平衡计算中可以相互抵消,即不考虑条间力作用。其他力均通过条块底面的中点,相应地 W,S,U 和 R'构造一个力多边形,如图 7-7b)所示,然后可以计算得到 R'。

base. Accordingly, a force polygon made up of the remaining forces, W, S, U, and R', is constructed [Figure 7-7b)], and R' is obtained.

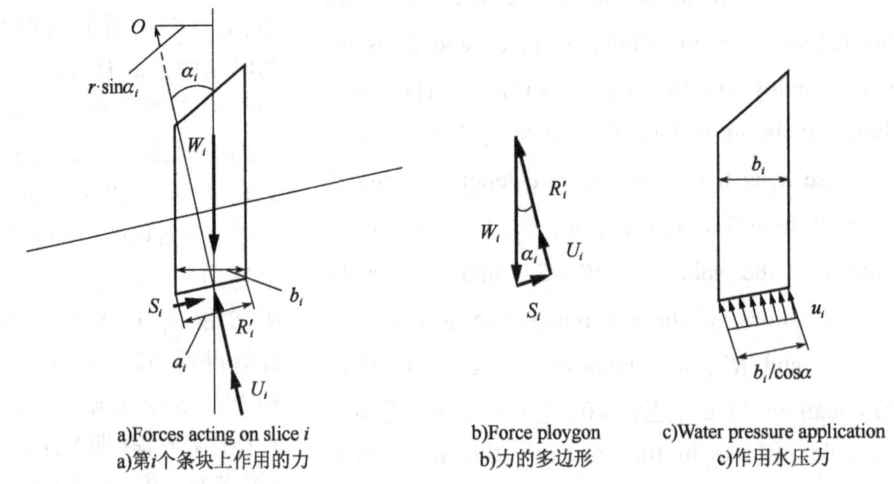

a)Forces acting on slice i
a)第i个条块上作用的力
b)Force ploygon
b)力的多边形
c)Water pressure application
c)作用水压力

Figure 7-7 Forces acting on slice i by the ordinary slice method
图 7-7 瑞典条分法中条块上的作用力

根据图 7-7b)有:

From Figure 7-7b):

$$R'_i = W_i \cdot \cos\alpha_i - U_i \tag{7-10}$$

因此,得到以下表达式。

Thus, we obtain the following expression.

$$F_s = \frac{M_{\text{resisting}}}{M_{\text{driving}}} = \frac{r[\sum c_i \cdot a_i + (R'_i \cdot \tan\varphi_i)]}{r\sum W_i \sin\alpha_i} = \frac{\sum\left[\dfrac{c_i \cdot b_i}{\cos\alpha_i} + (W_i \cdot \cos\alpha_i - U_i) \cdot \tan\varphi_i\right]}{\sum(W_i \cdot \sin\alpha_i)} \tag{7-11}$$

在式(7-11)中,r 是圆弧滑动面的半径;条块底部长度为 $a_i = b_i/\cos\alpha_i$;条块底部的水压力为 $U_i = u_i \cdot b_i/\cos\alpha_i$,如图 7-7c)所示。

In this equation, r is the radius of the sliding circle, and the base length of the slice is given by $a_i = b_i/\cos\alpha_i$. The resultant water pressure on the base length is $U_i = u_i \cdot b_i/\cos\alpha_i$, as seen in Figure 7-7c).

大量的计算结果表明,瑞典条分法计算的安全系数比其他精确解的值小约 10% 至 15%,在水压较高或坡度较平缓的情况下,安全系数可能会更低(低至 50%)。

It has been reported that this simple method provides an approximately 10% to 15% smaller (safer) than that obtained as the rigorous solution for ordinary slopes. In cases of higher water pressures and flatter slopes, the F_s value can be lower (as small as 50%) than these values.

7.3.2 临界滑面及 $F_{s\,min}$ 的求法

7.3.2 Calculation of critical slip surface and $F_{s\,min}$

上述 F_s 的计算只对某一具体的滑动面,土质边坡中可以假

The above calculation of F_s applies only to a specific sliding surface. Many such sliding surfaces can be assumed in soil slope calculations. Among all the assumed sliding surfaces, there must be a certain sliding surface whose corresponding F_s value is the minimum. The sliding surface corresponding to $F_{s\,min}$ is the most dangerous sliding surface and is called the critical sliding surface. In stability analyses of soil slopes, the stability of a soil slope be evaluated correctly only when the critical sliding surface and the corresponding $F_{s\,min}$ value are determined. Therefore, determining the critical slip surface and evaluating its factor of safety are the key aspects of the analysis.

设存在很多这样的滑动面,假定滑动面中必有一个滑动面的 F_s 值最小。$F_{s\,min}$ 对应的滑动面是最危险的滑动面,称为临界滑动面。在土质边坡稳定性分析中,只有找到了临界滑动面与相应的 $F_{s\,min}$ 值,才能正确地评价土质边坡是否稳定。所以如何寻求临界滑动面,并计算其安全系数是分析中的关键。

Fellenius' research on a simple soil slope with homogeneous clay ($\varphi = 0$) shows that the center position can be determined according to the slope gradient ($1:m$), as shown in Figure 7-8, i. e., there are different angles β_1 and β_2 corresponding to different slope angles, as shown in Table 7-1. β_1 and β_2 intersect at the critical circle center.

费伦纽斯(Fellenius)对简单均质黏性土边坡($\varphi = 0$)的研究表明,圆心位置可根据边坡的坡度($1:m$)来确定。如图 7-8 所示,不同的坡角有不同的 β_1、β_2 角,如表 7-1 所示。β_1、β_2 角的交点即为临界圆心的位置 O。

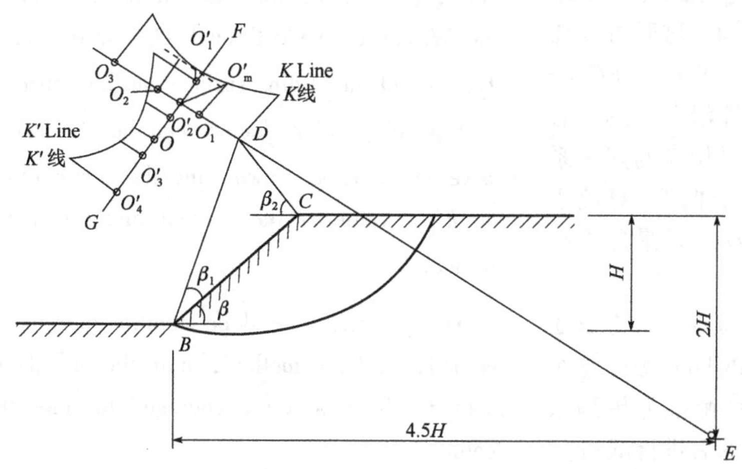

Figure 7-8 Determination of the circle center of the most dangerous sliding surface

图 7-8 确定最危险滑动面圆心的位置

Table 7-1 Determination of β_1 and β_2

表 7-1 β_1、β_2 的确定

坡度		β_1	β_2	坡度		β_1	β_2
1 : m	β			1 : m	β		
1 : 0.50	63°26′	29°30′	40°	1 : 1.75	29°45′	26°	35°
1 : 0.75	53°08′	29°00′	39°	1 : 2.00	26°34′	25°	35°
1 : 1.00	45°00′	28°	37°	1 : 3.00	18°26′	25°	35°
1 : 1.25	38°40′	27°	35°30′	1 : 5.00	11°19′	25°	37°
1 : 1.50	33°41′	26°	35°				

对于$\varphi>0$的黏性土,费伦纽斯(Fellenius)认为随着φ值增大,临界圆心位置向上且向外移动,可以过O点作直线OE(E点距坡脚B点的水平距离为$4.5H$,H为土质边坡的高度)来表示临界圆心的轨迹线,如图7-8所示。计算时从D点向外延伸几个试算圆心O_1、O_2,分别求得其相应的滑动稳定安全系数F_{s1}、F_{s2}……,绘制F_s值曲线可得最小安全系数值$F_{s\min}$,其相应的圆心即为最危险滑动面圆心。

For a clayey soil and $\varphi>0$, Fellenius postulated that as the value of φ increases, the position of the critical circle center moves upward and outward. A straight line OE (for a horizontal distance from point E to point B of the slope toe of $4.5\ H$, where H is the height of the soil slope) can be drawn to represent the trajectory line of the critical circle center, as shown in Figure 7-8. Extending several points from point D and calculating the centers of circles O_1, O_2, etc., the corresponding safety factors F_{s1}, F_{s2}, etc. can be obtained. The minimum safety factor $F_{s\min}$ can then be obtained by plotting the F_s value curve, and the corresponding center of the circle is the center of the critical sliding surface.

实际上土坡的最危险滑动面圆心位置有时并不一定在ED的延长线上,而可能在其左右附近,因此圆心O_m可能并不是最危险滑动面的圆心,这时可以通过O_m点作DE线的垂线FG,在FG上取几个试算滑动面的圆心O'_1、O'_2等,并求得相应的安全系数。再绘制F_s值曲线,对应于$F'_{s\min}$值的圆心O才是最危险滑动面的圆心。

The center of the critical sliding surface is sometimes not on the extended line of ED but rather may be on its left or right. Therefore, O_m may not be the center of the critical sliding surface. In this case, a vertical line FG through the line DE can be made through O_m; several circle centers O'_1, O'_2, etc. of the sliding surface can be drawn on FG, and the corresponding safety factors can be obtained. The F_s value curve can then be drawn, and the center O corresponding to the $F'_{s\min}$ value is the critical circle center of the sliding surface.

此外,边坡稳定分析方法还包括毕肖甫(Bishop)法、简布(Janbu)法、不平衡推力传递法等,建议读者在课后进行试用。

Other methods for slope stability analysis include Bishop's method, Janbu's method, and the imbalance thrust force method. Readers are encouraged to refer them for further reading.

例题 7-2

采用条分法计算图7-9所示边坡的安全系数。峰值强度参数$c'=10$ kPa,$\varphi'=29°$,地下水位以上和以下的土体重度均为20 kN/m^3。

Example 7-2

Using the Fellenius method of slices, determine the factor of safety, in terms of effective stress, of the slope shown in Figure 7-9 for the given failure surface using peak strength parameters $c'=10$ kPa and $\varphi'=29°$. The unit weight of the soil both above and below the water table is 20 kN/m^3.

解:

(1)安全系数由式(7-11)给出,土体被分成1.5 m宽的条块,每一条块的重量为$W=\gamma bh=20\times1.5\times h=30h$ kN/m。

Solution

(1)The factor of safety is given by Equation (7-11). The soil mass is divided into 1.5-m-wide slices. The weight (W) of each slice is given by $W=\gamma bh=20\times1.5\times h=30h$ kN/m.

(2) The height h for each slice is set off below the center of the base, and the normal and tangential components $h\cos\alpha$ and $h\sin\alpha$, respectively, are determined graphically, as shown in Figure 7-9.

(2)每一条块的高度 h 标注在底部中心,法向分量和切向分量($h\cos\alpha$)和($h\sin\alpha$)分别用图形标识确定,如图 7-9 所示。

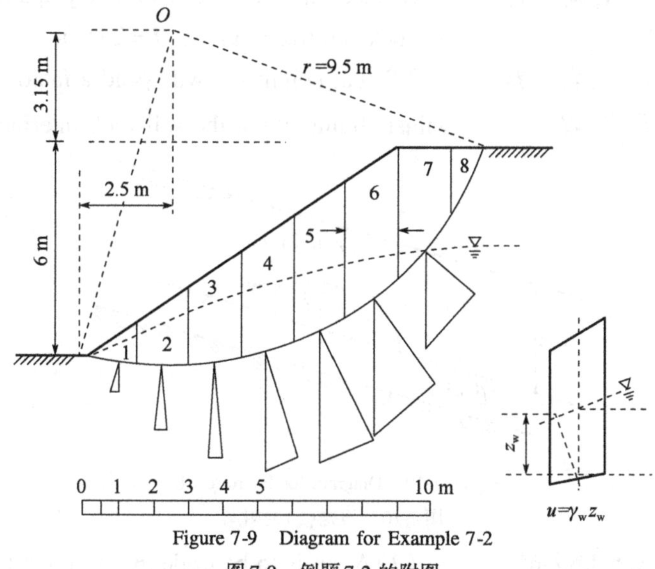

Figure 7-9　Diagram for Example 7-2

图 7-9　例题 7-2 的附图

Then:

那么:

$$W\cos\alpha = 30h\cos\alpha$$

The pore water pressure at the center of the base of each slice is taken to be $\gamma_w z_w$, where z_w is the vertical distance of the center point below the water table.

各条块底部中心位置的孔隙水压力为 $\gamma_w z_w$,其中 z_w 为地下水位以下中心点的垂直距离。

The arc length (L_α) is calculated as 14. 35 m. The results are provided in Table 7-2.

计算得出的弧长(L_α)为 14. 35 m,结果见表 7-2。

$$\sum W\sin\alpha = 30 \times 8.45 = 253.5\,(\text{kN/m})$$

$$\sum (W\cos\alpha - ul) = 525 - 132 = 393\,(\text{kN/m})$$

$$F_s = \frac{c'L_\alpha + \tan\varphi' \sum (W\cos\alpha - ul)}{\sum W\sin\alpha} = \frac{143.5 + 218}{254} = 1.426$$

Table 7-2　Calculation procedure for Example 7-2

表 7-2　例题 7-2 的计算步骤

Slice No. 条块编号	$h\cos\alpha$(m)	$h\sin\alpha$(m)	u(kN/m^2)	l(m)	ul(kN/m)
1	0.75	-0.15	5.9	1.55	9.1
2	1.80	-0.10	11.8	1.50	17.7
3	2.70	0.40	16.2	1.55	25.1
4	3.25	1.00	18.1	1.60	19.0
5	3.45	1.75	17.1	1.70	29.1
6	3.10	2.35	11.3	1.95	22.0
7	1.90	2.25	0	2.35	0
8	0.55	0.95	0	2.15	0
	17.50	8.45		14.35	132.0

习　题

(1)考虑图 7-10 所示的无限长边坡。

①给定 $H=2.4$ m,确定岩土界面抗滑安全系数。

②在什么高度 H 下,岩土界面滑动的安全系数 F_s 为 2?

Exercises

(1) Consider the infinite slope shown in the Figure 7-10.

①Determine the factor of safety against sliding along the soil-rock interface, given $H=2.4$ m.

②What height H, will yield a factor of safety, F_s, of 2 against sliding along the soil-rock interface?

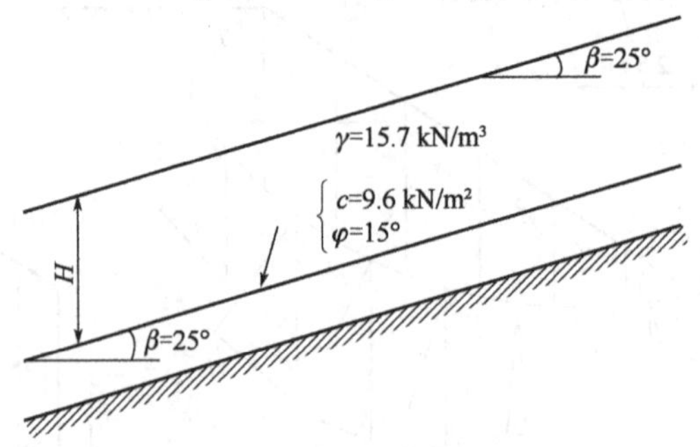

Figure 7-10　Diagram for Exercise (1)

图 7-10　习题(1)的附图

(2)要在 $\gamma=16.5$ kN/m³, $c=29$ kN/m² 和 $\varphi=15°$ 土体上设计与水平面成 45°角的切坡。边坡切深为多少,安全系数 F_s 为 3?

(2) A cut is to be made in a soil that has $\gamma=16.5$ kN/m³, $c=29$ kN/m², and $\varphi=15°$. The side of the cut slope makes an angle of 45° with the horizontal. What depth of the cut slope will have a factor of safety (F_s) of 3?

(3)如图 7-11 所示,一块岩石位于斜坡上。计算该块体的抗滑安全系数。如果水库中的边坡和岩石完全淹没在水中,新的安全系数是什么?对于这两种情况,假定块体底部的抗剪强度指标为:摩擦角 32°和黏聚力 100 kPa。垂直于页面方向的尺寸为 3m,岩石的密度为 2400 kg/m³。

(3) A block of rock lies on a slope as shown in Figure 7-11. Calculate the factor of safety against sliding for this block. If the slope and rock are completely submerged in water in a reservoir, what will be the new factor of safety? For both cases, assume that the shear strength at the base of the block is governed by a friction angle of 32° and a cohesion of 100 kPa. The width of the block into the page is 3 m, and the density of the rock is 2400 kg/m³.

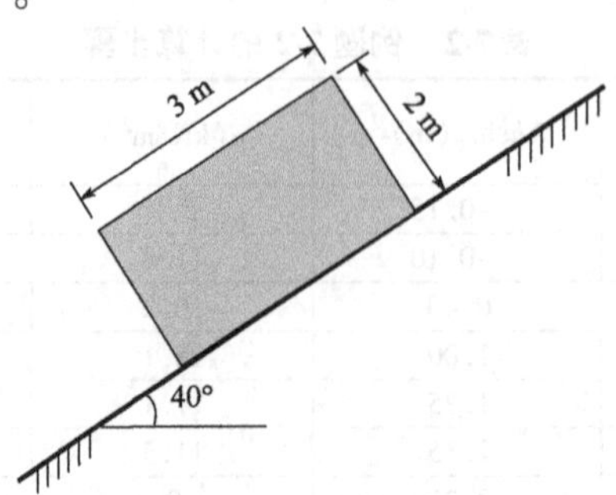

Figure 7-11　Diagram for Exercise(3)

图 7-11　习题(3)的附图

Foundation Bearing Capacity 8 Chapter
地基承载力

Problems in building foundation often lead to types of damage: one is that the foundation soil exhibits compression deformation under the building load, which leads to excessive settlement or differential settlement, causing the superstructure to incline and crack; the other is that the building load is extremely large and exceeds the bearing capacity of the soil under the foundation, resulting in a shear failure of the foundation. The types of shear failure that can occur are generally classified as general, local, and punching shear failures. This chapter focuses on the type of shear failure caused by the insufficient bearing capacity of a foundation and the evaluation of the bearing capacity of the foundation.

建筑物地基的破坏有两种情况:一是地基土在建筑物荷载作用下产生压缩变形,引起基础过大的沉降量或沉降差,使上部结构倾斜、开裂以致毁坏;二是由于建筑物的荷载过大,超过了基础下土体持力层所能承受荷载的能力而使地基产生剪切破坏。剪切破坏的形式一般分为整体剪切破坏、局部剪切破坏和冲剪破坏。本章主要讨论由于地基承载力不足引起的剪切破坏及地基承载力的计算。

8.1 Instability form and development process of foundation soil

8.1 地基的失稳形式和过程

The three types of foundation shear failures mentioned above are illustrated in Figure 8-1 and are described here with reference to a strip footing. In the case of general shear failure [Figure 8-1a)], continuous failure surfaces develop between the edges of the footing and the ground surface. When the load on the foundation is small, the foundation pressure changes approximately linearly with the settlement, which corresponds to the elastic deformation stage (section *OA* of the curve). As the pressure increases toward the value q_f, a state of plastic equilibrium is reached initially in

图 8-1 为地基破坏的三种形式,图 8-1a)表示整体剪切破坏,当基础上的荷载较小时,基底压力与沉降量呈近似线性关系,此时属于弹性变形阶段(曲线中 *OA* 段);随着荷载增大到某一值时,基础边缘处的土体开始出现剪切破坏(或称塑性破坏),剪切破坏区随着荷载的增大而扩大,此时基底压力与沉降

量呈曲线特征,此时属于弹塑性变形阶段(AK段);如基础上的荷载继续增加,剪切破坏区将继续扩展,此时说明基础上的荷载已经达到地基土的最大承载能力,地基濒于破坏,一旦荷载略增,基础将急剧下沉或突然倾倒,同时基础两侧的地面向上隆起而破坏,此时属于塑性破坏阶段。地基土开始出现剪切破坏时,地基所承受的基底压力称为临塑荷载,以 P_{cr} 表示。当基础急剧下沉、地基中出现连续滑动面(连续塑性区),地基土沿此滑动面从基底一侧或两侧大量挤出,整个地基将失去稳定时,地基所承受的基底最大压力称为极限荷载或地基的极限承载力,以 P_k 表示。

the soil around the edges of the footing, which then gradually spreads downward and outward, corresponding to the stage of elastic-plastic deformation (section AK). Ultimately, the state of plastic equilibrium is fully developed throughout the soil above the failure surfaces. Heaving of the ground surface occurs on both side of the footing, although the final slip movement occurs only on one side, accompanied by the tilting of the footing. In general, when a shear failure of the foundation soil begins, the pressure at the base of the foundation is called the critical plastic pressure, expressed as P_{cr}. When the foundation sinks rapidly and there is a continuous sliding surface (continuous plastic zone) in the foundation, the foundation soil is squeezed out from one side or both sides of the foundation along the sliding surface, and the entire foundation loses stability. In this case, the maximum pressure on the foundation is called the ultimate pressure or the ultimate bearing capacity of the foundation soil, expressed as P_k.

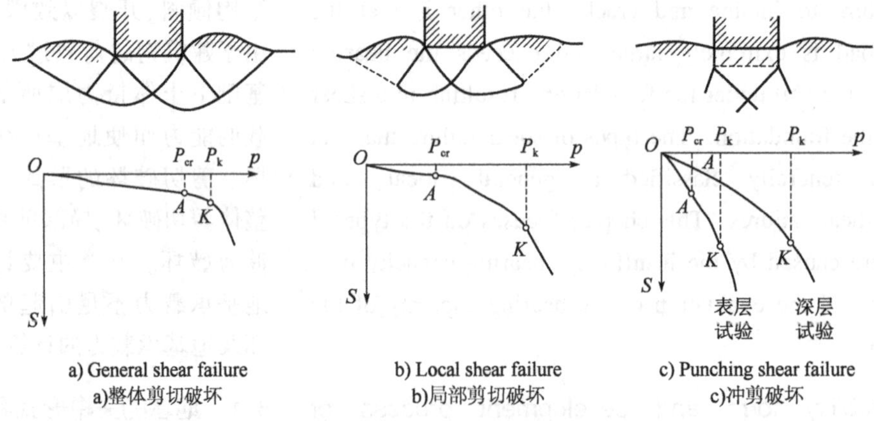

a) General shear failure
a)整体剪切破坏

b) Local shear failure
b)局部剪切破坏

c) Punching shear failure
c)冲剪破坏

Figure 8-1 Modes of foundation shear failure

图 8-1 地基的破坏模式

为保证建筑物的正常安全使用,常把基底压力限制在某一容许的承载力之内,该容许承载力以[P]表示。它等于地基的极限承载力 P_k 除以安全系数 K,即[P] = P_k/K。它表示地基的承载力应有一定的安全储备。因此,只要基底压力 P 小于或等于地基容许承载力,即 $P \leqslant [P]$,就认为基础是安全可靠的。

To ensure the safe use of buildings, the base pressure is usually limited to an allowable bearing capacity, expressed as [P], which is equal to the ultimate bearing capacity P_k of the foundation divided by a safety factor K, i.e., [P] = P_k/K. This means that the bearing capacity of the foundation should have a safety reserve. Therefore, when the base pressure P is less than or equal to the allowable bearing capacity of the foundation soil, that is, $P \leqslant [P]$, the foundation is considered to be safe.

The process of local shear failure [Figure 8-1b)] is similar to that of general shear failure. The shear failure starts from the foundation edge. With increasing load, the shear failure zone expands correspondingly. In the mode of local shear failure, there is significant compression in the soil under the footing and only a partial development of the state of plastic equilibrium. The failure surfaces, therefore, do not reach the ground surface, and only a slight heaving occurs. The relationship between pressure and settlement changes nonlinearly, and there is no obvious turning phenomenon when failure is reached. In this case, the stress corresponding to the point with a significant slope on the pressure-settlement curve is considered the ultimate bearing capacity of the foundation.

局部剪切破坏的过程与整体剪切破坏相似,其剪切破坏也从基础边缘下开始,随着荷载的增大,剪切破坏区相应地扩展。当荷载到某一值后,基础两侧的地面微微隆起呈现出破坏的特征,但剪切破坏区仅为地基内部的某一区域,并不能形成延伸至地面的连续滑动面,如图 8-1b)所示。局部剪切破坏时,其压力与沉降量的关系是非线性的,且达到破坏时无明显的转折现象。在这种情况下,压力-沉降曲线上斜率较大的点所对应的应力即为地基的极限承载力。

Punching shear failure [Figure 8-1c)] occurs when the compression of the soil under a footing is accompanied by shearing in the vertical direction around the edges of the footing. There is no heaving of the ground surface away from the edges and no tilting of the footing. The pressure-settlement curve of punching shear failure is similar to that of local shear failure, and there is no obvious turning phenomenon.

冲剪破坏[图 8-1c)]时,基底土体会产生压缩,同时基底边缘的土体会受到垂直方向的剪切。在这种情况下,地基边缘土体不会发生隆起,地基也不会倾斜。冲剪破坏的压力与沉降量关系为曲线特征且无明显的转折现象,类似局部剪切破坏的情况。

The shear failure mode of a foundation is related to the compression properties of the soil. In general, for hard or compacted soil, the failure mode will be general shear failure, whereas for soft soil, the failure mode will be either local or punching shear failure. The commonly used formulas for the bearing capacity of a foundation are derived for the general shear failure mode.

地基剪切破坏的形式与土的压缩性质有关。坚硬或紧密的土发生整体剪切破坏;松软土发生局部剪切或冲剪破坏。地基承载力公式均为整体剪切破坏条件下得到的。

8.2 Permissible bearing capacity of foundations

8.2 地基的容许承载力

In general, the failure of a base begins from the edge of the footing. At small load levels, the soil remains elastic. As the load reaches a certain level, soils adjacent to the footing reach a limit equilibrium condition first, and the base of the footing reaches this condition eventually. The corresponding load is defined as the critical edge pressure.

一般情况下,地基破坏从基础边缘开始。荷载较小时,土体保持弹性。当荷载达到一定值时,基础边缘土体首先达到极限平衡状态,地基发生屈服,相应的载荷定义为临塑荷载。

根据塑性变形区的深度确定地基承载力的方法是将地基中的剪切破坏区限制在一定范围内,此时地基土承受的基底压力为容许承载力,这是一个弹塑性混合课题。本节将介绍均布荷载作用下条形基础容许承载力的近似计算方法。

The goal of the method for determining the bearing capacity of a foundation based on the depth of the plastic deformation zone is to limit the shear failure zone in the foundation to within a certain range. Depending on the corresponding bearing capacity of the foundation soil, the pressure is the allowable bearing capacity. This is an elastic-plastic mixed problem. This section presents an approximate calculation method for the allowable bearing capacity of a strip foundation under a uniform load.

若条形基础的宽度为 B,埋置深度为 d,其底面上作用着竖向均布压力 p,如图 8-2 所示。根据弹性理论,地基中任一点 M 由条形均布压力所引起的大小主应力为

The strip footing of width B and embedment depth d, shown in Figure 8-2, is subjected to a uniform load p. The maximum and minimum principal stresses on an arbitrary element M are given by:

$$\left.\begin{matrix}\sigma'_1\\ \sigma'_3\end{matrix}\right\} = \frac{p-\gamma d}{\pi}(\beta_0 \pm \sin\beta_0) \tag{8-1}$$

式中,β_0 为 M 点与基础两侧连线的夹角。

where β_0 is the angle at M with respect to two endpoints of the uniform load (°).

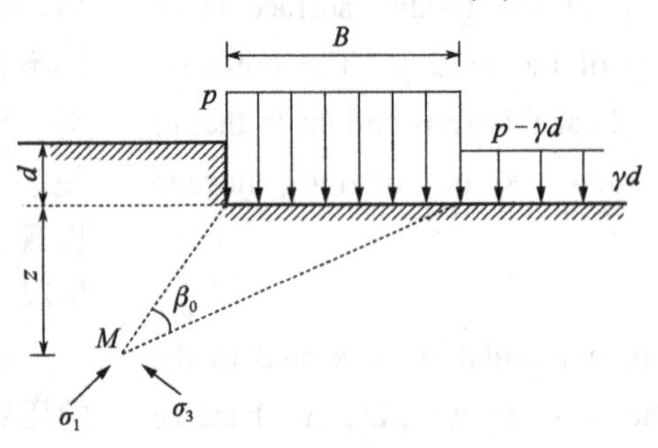

Figure 8-2 Principal stresses under a uniform load

图 8-2 均布荷载下的主应力分布

M 点的总应力为地基土的自重应力和附加应力的综合。在土体自重应力作用下,M 点的大主应力为垂直应力 $\sigma_{s1}=\gamma(d+z)$,小主应力为水平应力 $\sigma_{s3}=K_0\gamma(d+z)$,$K_0$为静止侧压力系数。由于 σ'_1和 σ_{s1} 的方向不一致,故不能代数相加。为简化计算,假定在极限平衡区内土的静止侧压力系数 $K_0=1$,则 $\sigma_{s1}=\sigma_{s3}$,均等于 $\gamma(d+z)$,即 M 点处土体自重应力相当于静水压力,

The gross stress on M is the sum of the gravity stress and the surcharge stress. Under the soil's self-weight, the maximum principal stress at point M is the vertical stress, $\sigma_{s1}=\gamma(d+z)$, while the minimum principal stress is the horizontal stress $\sigma'_{s3}=K_0\gamma(d+z)$, where K_0 is the static lateral pressure coefficient. Since the directions of σ'_1 and σ_{s1} are inconsistent, they cannot be added algebraically. For simplicity, it is assumed that the gravity stress field is regarded as a hydrostatic stress field, i. e., the lateral pressure coefficient is equal to 1.0, then $\sigma_{s1}=\sigma_{s3}$, are equal to $\gamma(d+z)$. The total stress on M is then given by:

任何方向的应力是相等的。因此,基底压力与土体自重应力在 M 点引起的大小主应力分别为:

$$\frac{\sigma_1}{\sigma_3} = \frac{p - \gamma d}{\pi}(\beta_0 \pm \sin\beta_0) + \gamma(d + z) \tag{8-2}$$

When the element M reaches the limit equilibrium condition, according to the Mohr-Coulomb criterion, the following principal stresses exist:

根据莫尔-库仑破坏准则,当 M 点达到极限平衡时,其大小主应力应满足下列关系:

$$\sigma_1 = \sigma_3 \cdot \tan^2\left(45° + \frac{\varphi}{2}\right) + 2c \cdot \tan\left(45° + \frac{\varphi}{2}\right) \tag{8-3}$$

Substituting Equation (8-2) into Equation (8-3), the depth of plastic zone is obtained as follows:

将式(8-2)代入式(8-3),得到塑性区深度:

$$z = \frac{P - \gamma d}{\gamma \pi}\left(\frac{\sin\beta_0}{\sin\varphi} - \beta_0\right) - \frac{c}{\gamma \cdot \tan\varphi} - d \tag{8-4}$$

Equation (8-4) is the boundary equation for the plastic zone, which reflects the relationship between z and β_0 in the plastic zone. If γ, c, φ, and p are known, the plastic zone boundary can be drawn.

式(8-4)是塑性区的边界方程,反映了塑性区中 z 和 β_0 之间的关系。如果已知 γ、c、φ 和 p,就可以绘制塑性区边界。

In practical applications, it is not necessary to know the boundary of the whole plastic zone but rather to know the maximum depth of the plastic zone under a certain base pressure. The maximum depth of the plastic zone can be obtained by differentiating Equation (8-4). Take the derivative of Equation (8-4) with respect to β_0, and set it value as 0.

在实际工程中,并不需要知道整个塑性区的边界,只需了解在一定基底压力作用下塑性区的最大深度。为求取塑性区的最大深度,将式(8-4)对 β_0 求导,并令其等于零,即:

$$\frac{dz}{d\beta_0} = \frac{p - \gamma d}{\gamma \pi}\left(\frac{\cos\beta_0}{\sin\varphi} - 1\right) = 0 \tag{8-5}$$

Then $\cos\beta_0 = \sin\varphi$, thus:

由此得 $\cos\beta_0 = \sin\varphi$,那么

$$\beta_0 = \frac{\pi}{2} - \varphi \tag{8-6}$$

Substituting Equation (8-6) into Equation (8-4) yields the following equation for the maximum depth:

将 β_0 代回式(8-4),则可得塑性区的最大深度为:

$$z_{\max} = \frac{p - \gamma d}{\gamma \pi}\left(\cot\varphi - \frac{\pi}{2} + \varphi\right) - \frac{c}{\gamma \cdot \tan\varphi} - d \tag{8-7}$$

If the allowable depth [z] of the plastic zone is specified, the stability of the foundation can be judged according to the following relationship:

如规定了塑性区深度的容许值[z],则可按下列关系式判别地基的稳定性:

如果 $z_{max} \leq [z]$,地基是稳定的;

If $z_{max} \leq [z]$, the foundation is stable.

如果 $z_{max} > [z]$,地基的稳定是无保证的。

If $z_{max} > [z]$, the stability of the foundation is not guaranteed.

根据经验,塑性区深度的容许值 $[z] = (1/4 \sim 1/3)B$,其中 B 为条形基础宽度。在实际工程中,可根据塑性区深度的容许值计算其可承受的基底压力并判别地基的稳定性。此时,式(8-7)改写为:

The allowable depth of the plastic zone $[z]$ is typically equal to 1/4th to 1/3rd of B, where B is the width of the strip foundation. In practical engineering applications, the stability of a foundation can be judged by calculating the foundation pressure with respect to the allowable depth of the plastic zone. Therefore, Equation (8-7) becomes:

$$p = \frac{\gamma \pi z_{max}}{\cot\varphi - \frac{\pi}{2} + \varphi} + \gamma d\left(1 + \frac{\pi}{\cot\varphi - \frac{\pi}{2} + \varphi}\right) + c\left(\frac{\pi \cdot \cot\varphi}{\cot\varphi - \frac{\pi}{2} + \varphi}\right) \tag{8-8}$$

若 $z_{max} = 0$,则塑性区深度为零。此时,地基承受的基底压力称为临塑荷载 P_{cr}。

If $z_{max} = 0$, the depth of the plastic zone is zero. Under this condition, the pressure that the foundation can bear is called the critical edge pressure P_{cr}.

$$P_{cr} = \gamma d\left(1 + \frac{\pi}{\cot\varphi - \frac{\pi}{2} + \varphi}\right) + c\left(\frac{\pi \cdot \cot\varphi}{\cot\varphi - \frac{\pi}{2} + \varphi}\right) \tag{8-9}$$

若 $z_{max} = B/4$,即塑性区最大深度限制在基础宽度的1/4,此时相应的地基容许承载力为:

If $z_{max} = B/4$, i. e. , if the maximum depth of the plastic zone is limited to 1/4th of the foundation width, the corresponding allowable bearing capacity of the foundation is given by:

$$[P] = P_{1/4} = \frac{\gamma \pi \cdot B}{4\left(\cot\varphi - \frac{\pi}{2} + \varphi\right)} + \gamma d\left(1 + \frac{\pi}{\cot\varphi - \frac{\pi}{2} + \varphi}\right) + c\left(\frac{\pi \cdot \cot\varphi}{\cot\varphi - \frac{\pi}{2} + \varphi}\right) \tag{8-10}$$

若 $z_{max} = B/3$,相应的地基容许承载力为:

If $z_{max} = B/3$, the corresponding allowable bearing capacity of foundation is:

$$[P] = P_{1/3} = \frac{\gamma \pi B}{3\left(\cot\varphi - \frac{\pi}{2} + \varphi\right)} + \gamma d\left(1 + \frac{\pi}{\cot\varphi - \frac{\pi}{2} + \varphi}\right) + c\left(\frac{\pi \cdot \cot\varphi}{\cot\varphi - \frac{\pi}{2} + \varphi}\right) \tag{8-11}$$

式(8-9)、式(8-10)和式(8-11)可用通式表示,即:

Equations (8-9), (8-10), and (8-11) can be expressed by the following general formula:

$$[P] = \frac{1}{2}\gamma B \cdot N_\gamma + \gamma d N_q + c N_c \tag{8-12}$$

式中,N_γ、N_q 和 N_c 为承载力系数,均为内摩擦角 φ 的函数,可查表8-1。

In this formula, N_γ, N_q, and N_c are bearing capacity coefficients that are functions of the internal friction angle φ. Values of these coefficients have been provided in Table 8-1.

$$N_c = \frac{\pi \cdot \cot\varphi}{\cot\varphi - \frac{\pi}{2} + \varphi}$$

$$N_q = 1 + N_c \cdot \tan\varphi$$

N_γ is related to the depth of the plastic zone z_{max}. When $z_{max} = 0$, $N_\gamma = 0$.

N_γ与塑性区最大深度z_{max}有关。

当$z_{max} = 0$时,$N_\gamma = 0$;

When $z_{max} = B/4$,

当$z_{max} = B/4$时,

$$N_\gamma = \frac{\pi}{2\left(\cot\varphi - \frac{\pi}{2} + \varphi\right)}$$

When $z_{max} = B/3$,

当$z_{max} = B/3$时,

$$N_\gamma = \frac{2\pi}{3\left(\cot\varphi - \frac{\pi}{2} + \varphi\right)}$$

Equation (8-10) is derived under the condition of a uniform load on a strip foundation. For the purpose of checking the stability of buildings during construction, values of the shear strength indices c and φ of the foundation soil are typically adopted based on the results of quick shear testing. In general, the allowable bearing capacity of the foundation should be $P_{1/4}$ or $P_{1/3}$ rather than the critical edge pressure P_{cr}. For a soft clay with a very small value of φ (e. g., less than 5°), the differences between P_{cr} and $P_{1/4}$ or $P_{1/3}$ are very small; thus, they can be used arbitrarily.

式(8-10)是根据条形基础均布荷载的情况导出的。为了检查施工期间建筑物的稳定性,应根据地基土快剪试验获得的c和φ进行计算。通常情况下,地基容许承载力为$P_{1/4}$或$P_{1/3}$,而不是临塑荷载P_{cr}。对于φ值很小(如$\varphi < 5°$)的软黏土,P_{cr}与$P_{1/4}$或$P_{1/3}$之间相差甚小,因此可任意使用。

The above formula applicable in the case of a homogeneous foundation. If there are different soil layers above and below the foundation, the first term in Equation (8-12) should be based on the unit weight of the soil under the base, and the second term should be based on the bulk density of the soil above the base. In addition, the bulk density of the soil below the groundwater level should be based on the submerged unit weight γ_b.

上述公式是在均质地基的情况下得到的,如果基底上下是不同的土层,则式(8-12)中第一项应采用基底下土的重度,而第二项应采用基底以上土的重度。另外,地下水位以下土的重度一律采用浮重度γ_b。

Table 8-1 Values of $N_{1/4}$, $N_{1/3}$, N_q, and N_c as functions of φ

表 8-1 不同φ值对应的$N_{1/4}$, $N_{1/3}$, N_q和N_c值

φ(°)	$N_{1/4}$	$N_{1/3}$	N_q	N_c
0	0	0	1.00	3.14
2	0.06	0.08	1.12	3.32
4	0.12	0.16	1.25	3.51
6	0.20	0.27	1.40	3.71
8	0.28	0.37	1.55	3.93
10	0.36	0.48	1.73	4.17

Continue

续上表

φ(°)	$N_{1/4}$	$N_{1/3}$	N_q	N_c
12	0.46	0.60	1.94	4.42
14	0.60	0.80	2.17	4.70
16	0.72	0.96	2.43	5.00
18	0.86	1.15	2.72	5.31
20	1.00	1.33	3.16	5.66
24	1.40	1.86	3.87	6.45
28	2.00	2.66	4.93	7.40
30	2.40	3.20	5.60	7.95
34	3.20	4.26	7.20	9.22
38	4.20	5.60	9.44	10.80
40	5.00	6.66	10.80	11.73
44	6.40	8.52	14.50	14.00
45	7.40	9.86	15.60	14.60

例题 8-1

某条形基础宽 $B=4$ m,埋置深度 $d=1$ m,地基土的天然重度 $\gamma=18$ kN/m³,饱和重度 $\gamma_s=20$ kN/m³,土的快剪强度指标 $c=5$ kN/m²、$\varphi=14°$。试求:

(1)地基的容许承载力 $P_{1/4}$,$P_{1/3}$;

(2)若地下水位上升至基础底面,承载力有何变化?

解:

(1)由表8-1查得 $\varphi=14°$时承载力系数 $N_{1/4}=0.60$、$N_{1/3}=0.80$,$N_q=2.17$,$N_c=4.7$。根据式$[P]=\frac{1}{2}\gamma B\cdot N_r+\gamma dN_q+cN_c$ 可得:

Example 8-1

A strip foundation with a width $B = 4$ m is buried to a depth $d = 1$ m in a base soil with a unit weight of $\gamma = 18$ kN/m³, a saturated unit weight $\gamma_s = 20$ kN/m³, $c = 5$ kN/m², and $\varphi = 14°$.

(1) Determine the parameters $P_{1/4}$ and $P_{1/3}$ of the allowable bearing capacity of the foundation.

(2) If the groundwater level rises to the bottom of the foundation, what will be the change in the bearing capacity?

Solution

(1) Based on Table 8-1, when $\varphi = 14°$, $N_{1/4} = 0.60$, $N_{1/3} = 0.80$, $N_q = 2.17$, and $N_c = 4.7$. Based on $[P] = \frac{1}{2}\gamma B\cdot N_r + \gamma dN_q + cN_c$:

$$P_{1/4}=\frac{1}{2}\gamma BN_{1/4}+\gamma dN_q+cN_c$$

$$=\frac{1}{2}\times18\times4\times0.60+18\times1\times2.17+5\times4.7=84.15\,(\text{kPa})$$

$$P_{1/3}=\frac{1}{2}\gamma BN_{1/3}+\gamma dN_q+cN_c$$

$$=\frac{1}{2}\times18\times4\times0.80+18\times1\times2.17+5\times4.7=91.36\,(\text{kPa})$$

(2) If the groundwater level rises to the bottom of the foundation, the bulk density of the soil below the groundwater level will be the buoyant unit weight $\gamma_b = \gamma_s - \gamma_w = 20 - 10 = 10\ \text{kN/m}^3$. If the c and φ values remain unchanged, the bearing capacity coefficient remains the same as above, hence:

(2)当地下水位上升至基础底面,则地下水位以下土的重度为浮重度 $\gamma_b = \gamma_s - \gamma_w = 20 - 10 = 10\ \text{kN/m}^3$。如 c,φ 值不变,则承载力系数同上,由此:

$$P_{1/4} = \frac{1}{2}\gamma_b BN_{1/4} + \gamma dN_q + cN_c$$

$$= \frac{1}{2} \times 10 \times 4 \times 0.60 + 18 \times 1 \times 2.17 + 5 \times 4.7 = 74.56(\text{kPa})$$

$$P_{1/3} = \frac{1}{2}\gamma_b BN_{1/3} + \gamma dN_q + cN_c$$

$$= \frac{1}{2} \times 10 \times 4 \times 0.80 + 18 \times 1 \times 2.17 + 5 \times 4.7 = 78.56(\text{kPa})$$

The calculation results show that the bearing capacity of the soil base will decrease if the groundwater level rises.

由计算结果表明,当地下水位上升时,地基的承载力将降低。

8.3 Prandtl's bearing capacity theory

8.3 普朗特承载力理论

Presently, the usual approach to determine the bearing capacity of a shallow foundation is limited to the general shear failure mode. The bearing capacity of a foundation for the general shear failure mode can be determined using several methods. This section presents the Prandtl formula, which serves as the foundation for an ultimate bearing capacity theory.

目前,浅基础的地基承载力是基于整体剪切破坏模式计算的。整体剪切破坏模式的地基极限承载力有多种方法确定,本节仅介绍作为极限承载力理论基础的普朗特(Prandtl)公式。

The foundation is assumed to be a soft, homogeneous, isotropic, and weightless medium. The shear strength parameters for the soil are c and φ, but the unit weight is assumed to be zero. The bottom of the foundation is smooth — that is, there is no friction between the bottom of the foundation and the soil.

假定地基为均匀软质且各向同性的无重量介质,即认为土体是一种 $\gamma = 0$,而有 c、φ 的材料;假定基础底面光滑,即基础底面与土之间没有摩擦力。

According to the elastoplastic limit equilibrium theory and the boundary conditions established by the above assumptions, the shape of the sliding surface is as shown in Figure 8-3. The area surrounded by the sliding surface is divided into five zones: one Zone Ⅰ, two Zones Ⅱ, and two Zone s Ⅲ. Since the bottom of the loading plate is assumed

根据弹塑性极限平衡理论和上述假定的边界条件,得出滑动面的形状如图 8-3 所示,滑动面所包围的区域分五个区:一个Ⅰ区,2 个Ⅱ区,2 个Ⅲ区。由于假设基础底面是光滑的,Ⅰ区中的竖向应力为大主应力,称为朗肯主动区,滑动面与水平面成 $(45° + \varphi/2)$。由于Ⅰ区的土楔 *ABC* 向下位移,把附近的土挤向两侧,使Ⅲ区中的土体 *ADF* 和 *BEG* 达到被动状态,称为朗肯被动区,滑动面与水平面成 $(45° - \varphi/2)$。主动区与被动区之间是一组对数螺线和一组辐射线组成的过渡区。对数螺线方程为 $r = r_0\exp(\theta\tan\varphi)$,若以 *A* 或 *B* 为极点,*AC* 或 *BC* 为 r_0,则可证明对数螺线分别与主动及被动区的滑动面相切。

to be smooth, the vertical stress in zone I is a maximum principal stress, and this becomes a Rankine active zone. The sliding surface is oriented at $(45° + \varphi/2)$ with respect to the horizontal plane. Because of the downward displacement of the soil wedge *ABC* in zone Ⅰ, the soil nearby is squeezed to both sides, so that segments *ADF* and *BEG* of the soil in Zone Ⅲ reach a passive Rankine state, and the sliding surface is oriented at $(45° - \varphi/2)$ with respect to the horizontal plane. Between the active region and the passive region is a transition region composed of a group of logarithmic spirals and a group of radiation lines. The logarithmic spiral equation is $r = r_0\exp(\theta\tan\varphi)$. If *A* or *B* is the pole and *AC* or *BC* is r_0, it can be proven that the logarithmic spiral is tangential to the sliding surfaces of the active and passive regions.

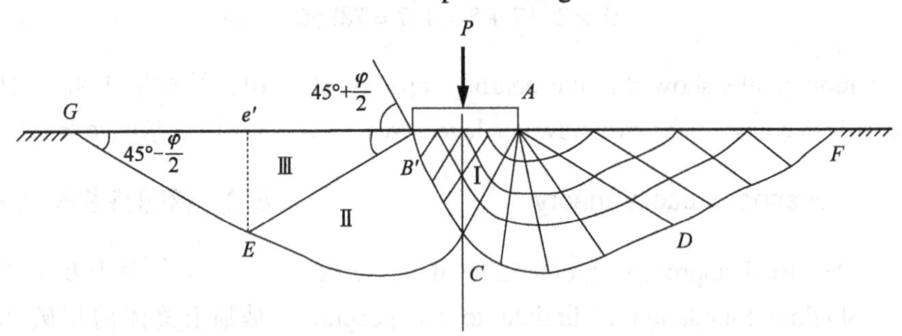

Figure 8-3 General failure under strip footing (Prandtl)

图 8-3 条形基础的整体破坏[普朗特(Prandtl)滑动面]

根据以上假定,普朗特导出极限承载力的理论解为:

Based on the above hypothesis, Prandtl derived the following theoretical solution for the ultimate bearing capacity:

$$P_u = cN_c \tag{8-13}$$

式中,N_c 为承载力系数,是 φ 的函数。

Where N_c is called the bearing capacity factor and is a function of φ.

$$N_c = \cot\varphi\left[\exp(\pi\tan\varphi)\tan^2\left(45° + \frac{\varphi}{2}\right) - 1\right] \tag{8-14}$$

若考虑基础埋深 d,则将基底平面以上的覆土以压力 $q = rd$ 代替,雷思诺(Reissner)(1924年)得出极限承载力的表达式为:

If the buried depth of the foundation is considered, the overburden above the base plane is replaced by the pressure $q = rd$. Reissner (1924) obtained the following expression for the ultimate bearing capacity:

$$P_u = cN_c + qN_q \tag{8-15}$$

Where N_c is calculated from Equation (8-14) and N_q is calculated as:

其中,N_c由式(8-14)计算得到,N_q计算方法为:

$$N_q = \exp(\pi\tan\varphi)\tan^2\left(45° + \frac{\varphi}{2}\right) \tag{8-16}$$

From the expressions of N_c and N_q, we obtain the following relationship:

从 N_c和 N_q的表达式可以看出:

$$N_c = (N_q - 1)\cot\varphi \tag{8-17}$$

According to Equation (8-15), when the foundation is placed on the surface of a cohesionless soil (c = 0 and d = 0), the bearing capacity of the foundation is equal to zero, which is unreasonable. This unreasonable situation is usually encountered since the soil is regarded to be a weightless medium. To overcome this defect, many scholars have developed revised approaches to determining the ultimate bearing capacity based on Prandtl's theory.

从式(8-15)可知,当地基置于无黏性土($c=0$)的表面上($d=0$)时,地基的极限承载力将等于零,这显然是不合理的。这一不合理的情况,主要是由于将土当作无重量介质造成的。为了弥补这一缺陷,许多学者根据普朗特理论提出了修正的极限承载力确定方法。

When the influence of the soil weight on the bearing capacity is considered, and for c = 0 and $q = \gamma d = 0$, the following expression for the ultimate bearing capacity can be stated:

当考虑土的重量对承载力的影响,且 $c=0$,$q=\gamma d=0$ 时,极限承载力的表达式如下:

$$P_u = \frac{1}{2}\gamma B N_\gamma \tag{8-18}$$

Where N_γ is a dimensionless bearing capacity factor. Vesic (1970) suggested the following approximate expression for N_γ:

式中:N_γ为无量纲承载力系数。Vesic(1970 年)提出了 N_γ的近似表达式:

$$N_\gamma \approx 2(N_q + 1)\tan\varphi \tag{8-19}$$

The approximation error associated with this expression is in the range of 5% ~10%, which is considered safe.

其误差在 5% ~10% 范围内,偏于安全。

For the case in which c, d, and γ are not zero, an equation for the ultimate bearing capacity is obtained by the superposition of Equations (8-15) and (8-18), resulting in:

对于 c(黏聚力)、d(距离地表距离)、γ 都不为零的情况,极限承载力公式系将式(8-15)和式(8-18)叠加,即为:

$$P_u = \gamma d N_q + c N_c + \frac{1}{2}\gamma B N_\gamma \tag{8-20}$$

The values of the bearing capacity coefficients N_q, N_c, and N_γ as functions of φ are provided in Table 8-2.

表 8-2 列出了承载力系数 N_q、N_c和 N_γ随 φ 变化的值。

Table 8-2 Values of N_q, N_c, and N_γ as functions of φ

表 8-2 不同 φ 值对应的 N_q、N_c、N_γ 值

φ	N_c	N_q	N_γ	φ	N_c	N_q	N_γ
0	5.14	1.00	0.00	26	22.25	11.85	12.54
1	5.38	1.09	0.07	27	23.04	13.20	14.47
2	5.63	1.20	0.15	28	25.80	14.72	16.72
3	5.90	1.31	0.24	29	27.86	16.44	19.34
4	6.19	1.43	0.34	30	30.14	18.40	22.40
5	6.49	1.57	0.45	31	32.67	20.63	25.99
6	6.81	1.72	0.57	32	35.49	23.18	30.22
7	7.16	1.88	0.71	33	38.64	26.09	35.19
8	7.53	2.06	0.86	34	42.16	29.44	41.06
9	7.92	2.25	1.03	35	46.12	33.30	48.03
10	8.33	2.47	1.22	36	50.59	37.75	56.31
11	8.80	2.71	1.44	37	55.63	42.92	66.19
12	9.28	2.97	1.69	38	61.35	48.93	78.03
13	9.81	3.26	1.97	39	67.87	55.96	92.25
14	10.37	3.59	2.29	40	75.31	64.20	109.41
15	10.93	3.94	2.65	41	83.86	73.90	130.22
16	11.63	4.34	3.06	42	93.71	85.38	155.55
17	12.34	4.77	3.53	43	105.11	99.02	186.54
18	13.10	5.26	4.07	44	118.37	115.31	224.64
19	13.93	5.80	4.68	45	133.88	134.88	271.76
20	14.83	6.40	5.39	46	152.10	158.51	330.35
21	15.82	7.07	6.20	47	173.64	187.21	403.67
22	16.88	7.82	7.18	48	199.26	222.31	496.01
23	18.05	8.66	8.20	49	229.93	265.51	613.16
24	19.32	9.60	9.44	50	266.89	319.07	762.89
25	20.72	10.66	10.88				

例题 8-2

某条形基础宽度 $B=6$ m，埋置深度 $d=1.5$ m，施加中心荷载 $F=1500$ kN/m³。地基土质均匀，重度 $\gamma=19$ kN/m³，土的抗剪强度指标 $c=10$ kN/m²、$\varphi=20°$。试验算地基的稳定性。安全系数 $K=2.5$。

解：

(1)基底压力为：

Example 8-2

Consider a strip foundation, with width $B=6$ m and embedded depth $d=1.5$ m, to which a centric load $F=1500$ kN/m³ is applied. The foundation soil is uniform; its bulk density is $\gamma=19$ kN/m³, and its shear strength index values are $c=10$ kN/m² and $\varphi=20°$. Calculate the stability of the foundation. The safety factor $K=2.5$.

Solution

(1) The contact pressure is:

$$P=\frac{F}{B}=\frac{1500}{6}=250(\text{kPa})$$

(2) For $\varphi=20°$, from Table 8-2, $N_c=14.83$, $N_q=6.40$, and $N_\gamma=5.39$. Inserting these values into Equation (8-20), the ultimate bearing capacity of the foundation is obtained as follows:

(2)由 $\varphi=20°$,查表8-2,得 $N_c=14.83$、$N_q=6.40$、$N_\gamma=5.39$,将上列各值代入式(8-20),得地基的极限承载力为:

$$P_u=\frac{1}{2}\times19\times6\times5.39+19\times1.5\times6.4+10\times14.83=637.93(\text{kPa})$$

For a safety factor $K=2.5$, the allowable bearing capacity of the foundation is:

安全系数 $K=2.5$,则地基容许承载力为:

$$[P]=\frac{P_u}{K}=\frac{637.93}{2.5}=255.17(\text{kPa})$$

Because $P<[P]$, the soil base is judged to be stable.

因为 $P<[P]$,所以地基是稳定的。

Exercises

习　题

(1) Consider a strip foundation with a bottom width $b=2.5$m buried at a depth $d=1.2$ m. The unit weight of the foundation soil is $\gamma=19$ kN/m^3; the saturated unit weight $\gamma_{sat}=19.8$ kN/m^3; the shear strength index values of the soil are $c=10$ kPa and $\varphi=10°$.

①Calculate the allowable bearing capacity values of the foundation $P_{1/4}$ and $P_{1/3}$.

② If the groundwater rises from very deep to the basement, what will be the change in the bearing capacity?

(2) A strip foundation of width $b=1.2$ m buried to a depth $d=2.0$ m is built on a homogeneous clay foundation with $\gamma=18$ kN/m^3, $\varphi=15°$, and $c=15$ kPa. Calculate the critical plastic load P_{cr} and boundary load $P_{1/4}$.

(3) A strip foundation with width $b=10$ m buried to a depth $d=2$ m is built on a homogeneous clay foundation with $\gamma=16.5$ kN/m^3, $\varphi=15°$, and $c=15$ kPa. If the groundwater level is at the bottom of the foundation ($\gamma'=8.7$ kN/m^3), what would be P_{cr} and $P_{1/4}$?

(1)某条形基础底宽 $b=2.5$ m,埋置深度 $d=1.2$ m。地基土重度 $\gamma=19$ kN/m^3,饱和重度 $\gamma_{sat}=19.8$ kN/m^3,土的抗剪强度指标 $c=10$ kPa,$\varphi=10°$,求解:

①地基的容许承载力 $P_{1/4}$ 和 $P_{1/3}$。

②若地下水由很深上升至基底面,承载力有何变化?

(2)某条形基础 $b=1.2$ m,$d=2.0$ m,建在均质的黏土地基上,黏土的 $\gamma=18$ kN/m^3,$\varphi=15°$,$c=15$ kPa,计算临塑荷载 P_{cr}和临界荷载 $P_{1/4}$。

(3)某条形基础宽度 $b=10$ m,埋置深度 $d=2.0$ m,建于均质黏土地基上,黏土的 $\gamma=16.5$ kN/m^3,$\varphi=15°$,$c=15$ kPa,若地下水位在基础底面处($\gamma'=8.7$ kN/m^3),P_{cr} 和 $P_{1/4}$ 各是多少?

Chapter 9 Engineering Geological Investigation 工程地质勘察

工程地质是土力学研究中最广泛的领域之一。任何建筑物在进行设计之前,都有必要对地基条件进行工程地质勘察,以确定各土层的工程特性和地基承载力。

One of the most widely studied fields of soil mechanics is engineering geology. Before the design of any building, it is necessary to carry out an engineering geological investigation of the foundation conditions to determine the engineering properties of each soil layer and the bearing capacity of the foundation.

9.1 勘察的基本步骤

9.1 Basic process of site exploration

在设计任何建筑物之前,都必须对场地进行勘察。勘察分三步进行。

The site must be explored before designing any structure. The exploration is performed in three steps.

(1)步骤一:初步踏勘现场。根据现有的初步信息,工程师需要评估项目的地形、位置、交通线路、排水条件、施工设备进场条件、附近结构物、水流、池塘及其他内容。根据需要对这些细节进行草图绘制、记录和拍照。可以使用手持探针检查地表土体条件,必要时还可挖掘一个或多个浅孔。

(1)Step 1: Preliminary site visit. With available preliminary information, engineers visit the site to assess the topography and the exact locations of the project site, driveways, drainage, access paths for construction equipment, nearby structures, water flow, ponds, and other features. These details are sketched, documented, and photographed as needed. Surface soil conditions can be checked using hand-held probes, and one or more shallow boreholes can be dug, if needed.

(2)步骤二:详细勘察现场。设计工程师对现场土体的分层情况进行评估,并采集土样对土体基本特性进行实验室测试,如重度、含水率、孔隙比、相对密度、

(2)Step 2: Detailed site exploration. Design engineers assess the stratification of the soil at the site and obtain samples for laboratory test of the soil's basic properties, including its unit weight, water content, void ratio, specific gravity, grain size distribution, Atterberg limits, compressibility, and shear strength parameters. Specific site exploration activities may include assessments of the following:

①Types of subsurface exploration equipment needed, possibly including geophysical methods

②Locations and depths of borings and/or test pits

③Types of field tests

④Locations and depths of field tests

⑤Locations and depths of soil sampling

⑥Groundwater table monitoring plan

Based on the exploration plan, various site exploration activities, including geophysical testing, boring, field tests, and sampling, are carried out.

(3) Step 3: Laboratory tests. Disturbed and undisturbed soil samples are brought to laboratories, and necessary tests, including those needed to determine the unit weight, water content, specific gravity, gradation, Atterberg limits, compaction, permeability, coefficient of consolidation, and shear strength, are conducted.

粒径分布、界限含水率、压缩性和抗剪强度参数等。具体的勘察内容包括:

①地下勘察设备类型,包括地球物理方法;

②钻孔或探坑的位置和深度;

③现场试验类型;

④现场试验的位置和深度;

⑤土体取样的位置和深度;

⑥地下水位监测方案

根据勘察计划,进行各种现场勘察活动,包括地球物理勘探、钻探、现场试验和取样。

(3)步骤三:室内试验。对扰动土样和原状土样进行室内试验,测定重度、含水率、相对密度、颗粒级配、界限含水率、压实度、渗透性、固结系数和剪切强度等指标。

9.2 In situ tests

9.2.1 Load testing

Load testing involves applying a static load to a rigid bearing plate of a certain size on site and measuring the bearing capacity and deformation characteristics of a natural soil base, single pile, or composite foundation within the main influence range of stress under the bearing plate. Depending on the shape and depth of the bearing plate, the test can be characterized as a shallow plate, a deep plate, or a spiral plate loading test. The equipment typically used in a plate loading test includes a pressure plate, loading system, reaction force system, and measuring system, as shown in Figure 9-1. A weight, a ground anchor, or a combination of a weight and a ground anchor is combined with a beam frame to form a stable reaction system. A displacement (settlement) measurement system includes a supporting column, datum beam, and displacement measuring elements (displacement sensors, dial indicators, etc.).

9.2 原位试验

9.2.1 载荷试验

载荷试验是指在现场对一定尺寸的刚性承载板施加静荷载,并测量承载板下应力主要影响范围内的天然地基、单桩或复合地基的承载能力和变形特性。根据承载板的形状和深度,试验可分为浅板、深板或螺旋板载荷试验。承载板载荷试验的设备包括承载板、加载系统、反力系统和测量系统,如图 9-1 所示。砝码、地锚或砝码与地锚的组合和梁架相结合,形成一个稳定的反力系统。位移(沉降)测量系统包括支撑柱、基准梁和位移测量元件(位移传感器、刻度盘指示器等)。

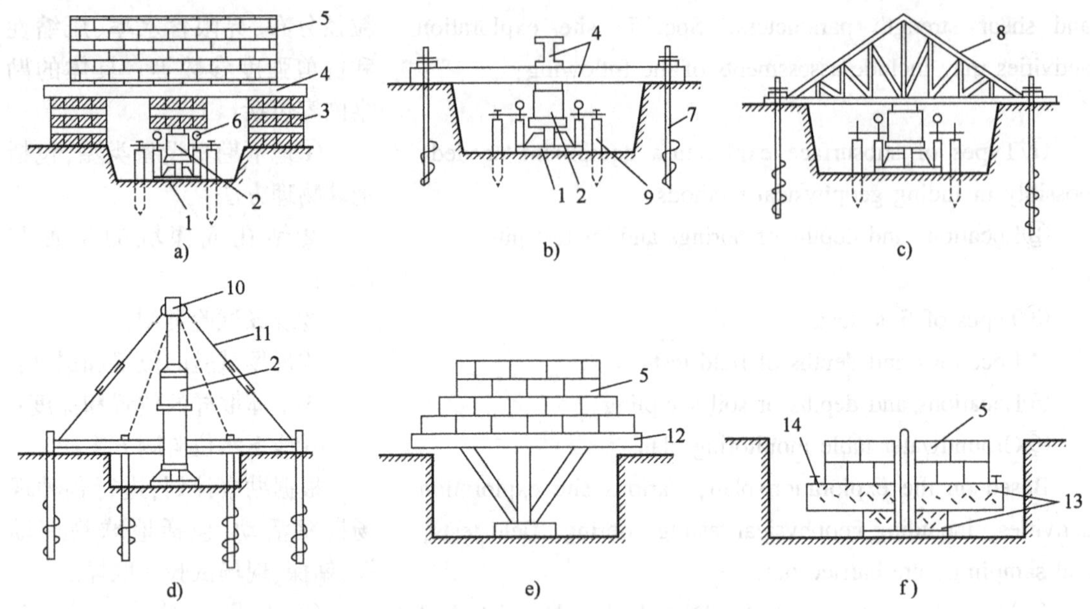

Figure 9-1 Common arrangement of loading test

图9-1 常见的载荷试验反力与加载布置方式

1-Pressure plate; 2-Jack; 3-Wooden crib; 4-Steel beam; 5-Steel ingot; 6-Dial indicator; 7-Ground anchor; 8-Truss; 9-Pillar; 10-Component cap; 11-Pull rod; 12-Load combination; 13-Concrete plate; 14-Test point

1-承压板;2-千斤顶;3-木垛;4-钢梁;5-钢锭;6-百分表;7-地锚;8-桁架;9-立柱;10-分力帽;11-拉杆;12-载荷台;13-混凝土板;14-测点

在平板载荷试验过程中,在地基持力层上放置一定尺寸和几何形状的刚性板(圆形或方形)。逐级增加荷载,测量每一级荷载下的地基沉降量,直至地基达到破坏标准,由此可得到荷载(p)-沉降(s)曲线(即p-s曲线)。

平板载荷试验获取的典型p-s曲线划分为三个阶段,如图9-2所示。①线性变形阶段:p-s曲线为直线段(线性关系),此段的最大压力P_{cr}称为临塑荷载,土体以压缩变形为主。②剪切变形阶段:当压力超过P_{cr},但小于极限荷载P_u时,压缩变形占比逐渐减少,剪切变形占比逐渐增加,在这一阶段,p-s线由直线变为曲线,曲线斜率逐渐增大。③破坏阶段:当荷载大于极限荷载P_u时,即使荷载不变,沉降也会急剧增加,不能满足稳定标准。

During plate load testing, rigid plates of a certain size and geometry (round or square) are placed on the bearing stratum of the soil base. The load is increased incrementally. A stable degree of settlement under each load is measured until the foundation failure criterion is reached. The measurements obtained are used to establish the load-settlement (p-s) curve for the soil.

A typical p-s curve derived from plate loading test results represents the following three stages of loading and deformation, as illustrated in Figure 9-2. ①The linear deformation stage, in which the p-s curve is a straight line. This stage terminates at the maximum pressure P_{cr} of this section, which is called the critical plastic pressure. The soil is mainly compressed in this stage. ②The shearing deformation stage, in which the pressure exceeds P_{cr} but is less than the limit pressure P_u; the proportion of compression in the total deformation decreases gradually, and the proportion of shear deformation increases gradually. In this stage, the p-s line changes from a straight line to a curve, and the slope of the curve increases gradually. ③The failure stage, in which the load exceeds the ultimate pressure P_u, such that even if the load remains unchanged, the settlement increases sharply and does not satisfy the stability criterion.

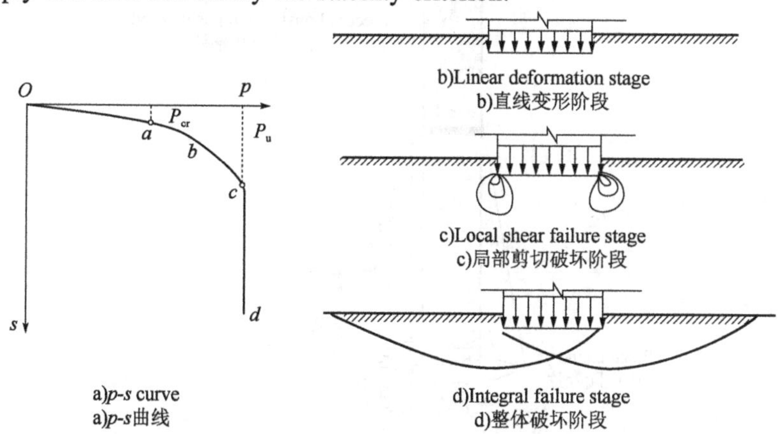

Figure 9-2 Three stages of *p-s* curve and the corresponding failure forms

图 9-2 变形曲线三阶段及相应地基破坏情况

9.2.2 Cone penetration test

In a cone penetration test (CPT), a cone penetrometer is pushed into a soil to measure its tip and frictional resistance, and in many cases, pore water pressure is generated and measured. Although cone penetration measurement systems were originally designed to operate mechanically (i. e., using a mechanical cone apparatus), most of the current systems measure the same parameters electrically (using an electrical cone apparatus). A typical cone penetrometer (piezocone) consists of a cone-shaped penetration tip, a frictional sleeve, and piezo elements for monitoring pore water pressure, as shown in Figure 9-3. In the traditional mechanical version of the CPT apparatus, the apex angle α_c is 60°, and the cone base area is 10 cm^2. The cone penetration resistance measured in CPT consists of two components: the cone resistance and the side wall resistance. A friction sleeve is mechanically isolated from the major shaft to monitor only the frictional resistance applied to the section using a load cell, and the pressure applied to the cone tip section is independently monitored using a load cell for the tip resistance. The cone is pushed into the soil statically by hydraulic power, and the cone tip resistance q_c, side friction f_s, and pore water pressure u are measured continuously and recorded electrically. Figure 9-4 shows an example of CPT (piezocone) data.

9.2.2 静力触探试验

静力触探试验是将一个圆锥贯入仪压入土中,测量土层对探头的贯入阻力和孔隙水压力。触探仪最初是机械式的(机械圆锥),但目前多采用电测触探仪,探头内嵌有压力传感器或电阻应变仪等电测传感器。常规触探仪由锥形贯入头、摩擦套筒和孔隙水压力计组成,如图 9-3 所示。探头锥角为 60°,锥底面积为 10 cm^2。静力触探试验的探头阻力由探头的锥尖阻力和侧壁阻力两部分组成。摩擦套筒与主轴隔离,用于监测侧壁阻力,锥尖阻力由应力传感器独立监测。触探仪由静力压入土中,连续测量锥尖阻力 q_c、侧摩阻力 f_s和孔隙水压力 u,测试数据的样图如图 9-4 所示。

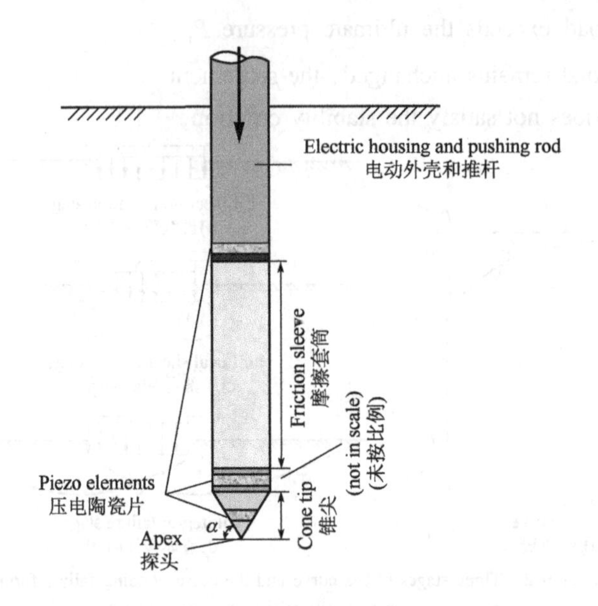

Figure 9-3 Typical cone penetrometer (piezocone)

图 9-3 典型的静力触探仪

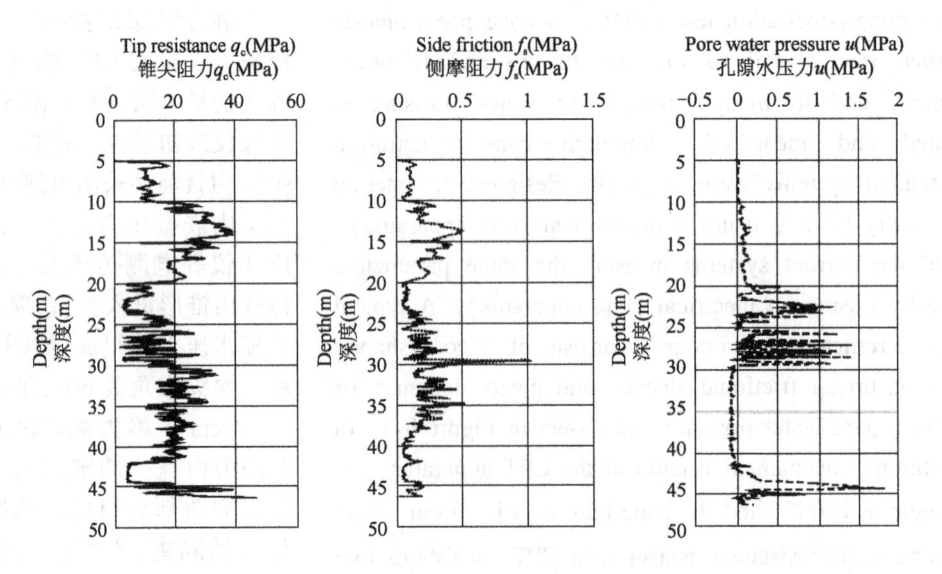

Figure 9-4 Example of CPT (piezocone) data

图 9-4 静力触探试验数据示例

静力触探试验测得的贯入阻力直接反映土体强度,因此结合静力触探试验的贯入阻力数据和土体载荷试验数据,可以建立土体贯入阻力和地基容许承载力之间的经验公式。该经验公式可以用来估计地基承载力。

The cone penetration resistance measured by CPT is a direct reflection of the soil strength. Therefore, an empirical correlation can be established between the penetration resistance of a given soil and the allowable bearing capacity of the natural foundation using CPT penetration resistance data and soil load test data. Such an empirical correlation can be used to estimate the foundation bearing capacity of the soil layer from the penetration resistance of the soil layer.

In the CPT data, a relatively high tip resistance q_c and a low pore water pressure u may be an indication of sandy soil layers. A relatively low q_c but higher f_s and u may indicate cohesive soils. Negative pore water pressure generation implies dilative soils, such as dense sands or highly overconsolidated clays. The friction ratio, R_f, defined as the ratio of the side friction to the tip resistance (f_s/q_c) is a useful indicator of soil types.

在静力触探试验中，高锥尖阻力 q_c 和低孔隙水压力 u 对应砂土；相对较低的 q_c 和较高的 f_s 与 u 对应黏土；负孔隙水压力对应膨胀土，如密砂或超固结黏土。摩阻比 R_f 定义为侧摩阻力与锥尖的端阻力之比(f_s/q_c)，是区分土体类型的有效指标。

9.2.3 Standard penetration test

The standard penetration test (SPT) has been the most widely performed test in geotechnical explorations for many years. The SPT involves driving a 51 mm outer-diameter (OD), 34.9 mm inner-diameter (ID) split-spoon sampler, as shown in Figure 9-5. A borehole is typically drilled to the desired depth with an auger drill surrounded by a casing. The SPT sampler is then lowered to the bottom of the borehole with a drilling rod. At the top of the drilling rod, a hammer with a 63.57 kg dead weight is dropped from a 76cm height to force the sampler into the soil. The first 15cm of penetration is considered to accomplish seating of the shoe on the undisturbed soil surface. The blow count for the next 30 cm of penetration is recorded as the standard penetration number or N value. At the end of the driving process, the sampler is raised to the ground level and is split open for specimen observation and collection for future laboratory testing. Figure 9-6 shows an example of a specimen recovered using an SPT split-spoon sampler. The specimen is evidently disturbed in this sampling procedure because of the rather thick sampler wall (7.95 mm) and should not be considered a truly undisturbed specimen. However, it can be used for index tests for parameters such as sieve analysis, specific gravity, and Atterberg limits.

9.2.3 标准贯入试验

标准贯入试验(SPT)在岩土工程勘察中的应用极为广泛。标准贯入试验的装置如图 9-5 所示，主要由一个 51 mm 外径(OD)和一个 34.9 mm 内径(ID)的对开式取样器组成。首先用钻探设备将钻孔钻到所需测试的土层，将标准贯入器套在钻杆上放至钻孔底；然后使 63.57 kg 的击锤从 76 cm 高度处自由下落，击打钻杆帽。通常在正式贯入之前，先将贯入器打入土中 15 cm(此时不计锤击数)，接着再打入土中 30 cm，此时所需的锤击数作为实测锤击数 N。试验结束后，观察土样形态并进行室内试验。图 9-6 为标准贯入试验对开式取样器的示例，由于采样器壁厚 7.95 mm，土样在试验过程中受到明显扰动，因此不应将其视为原状土，但该土样可用于颗粒级配、相对密度、界限含水率等试验。

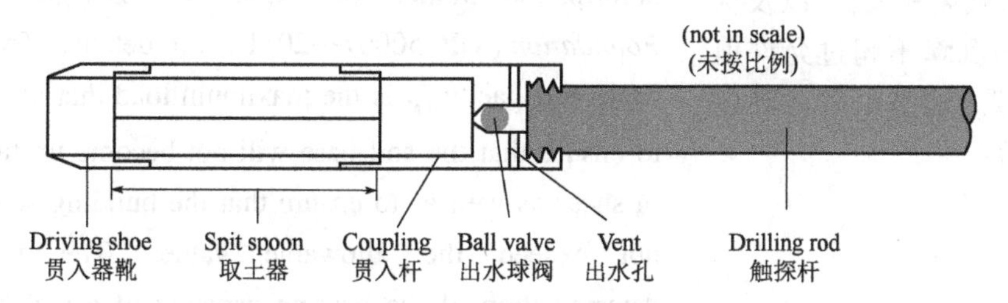

Figure 9-5 Schematic diagram of an SPT split-spoon sampler

图 9-5 标准贯入试验对开式取样器的结构示意图

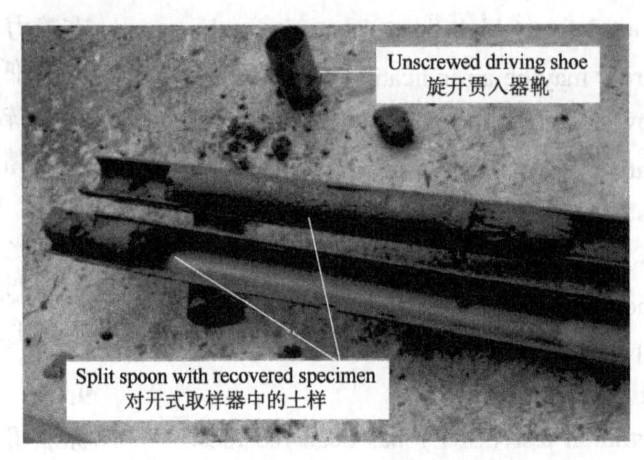

Figure 9-6 Specimen recovered in an SPT split-spoon sampler

图 9-6 标准贯入试验中对开式取样器中的土样

标准贯入击数 N 可直接反映土体抗剪强度,因此当土的标准贯入击数 N 已知时,可按照相关规范确定地基的容许承载力。

The standard penetration number N is directly related to the shear resistance of the soil; therefore, when N is known, the allowable bearing capacity of the foundation can be estimated based on correlations available in relevant codes.

9.3 地基承载力确定

9.3 Determination of foundation bearing capacity

为了满足地基稳定性的要求,设计时基础底面最大压力不得大于地基承载力。地基承载力包括特征值 f_a、容许值 f_{ak} 和极限值 P_u 等。特征值 f_{ak} 是从地基稳定的角度判断地基土体的承载能力,作为一个随机变量,它以概率理论为基础,由分项系数表达的实用极限状态设计法确定。同时需要验算地基变形是否超过容许值,详见《建筑地基基础设计规范》(GB 50007—2011)。容许承载力 f_{ak} 确保地基不发生剪切破坏和失稳,以及保证建筑物的沉降不超过允许值的最大荷载。地基承载力确定的方法包括:

To meet the requirements of soil base stability, the maximum pressure on a foundation must be controlled so that it does not exceed a certain value considered to be the foundation's bearing capacity. The bearing capacity of a soil foundation is characterized by, among other properties, its characteristic value f_a, allowable value f_{ak}, and ultimate value P_u. The characteristic value f_a of the foundation bearing capacity refers to the bearing capacity of the foundation with guaranteed reliability. As a random variable, it is based on the probability theory and determined by the practical limit state design method expressed by partial coefficients. At the same time, it is necessary to check that the foundation deformation does not exceed the allowable deformation value. See *Code for Design of Building Foundation* (GB 50007—2011) for details. The allowable bearing capacity f_{ak} is the maximum load that can be allowed to ensure that the soil base will not become unstable and fail in shear as well as to ensure that the building settlement will not exceed the allowable value. The methods for determination of the bearing capacity of a soil base include the following.

(1) In situ testing methods can be used to determine the bearing capacity of a foundation through on-site tests, including loading, static penetration, standard penetration, and pressuremeter tests. The loading test method is the most reliable in situ test method. The principle of determination of the bearing capacity of a soil base from a loading test is illustrated in Figure 9-7.

(1)原位试验。通过现场试验确定地基承载力,包括载荷试验、静力触探试验、标准贯入试验和压力计试验等。其中载荷试验是最可靠的,通过荷载试验确定地基承载力的原理如图9-7所示。

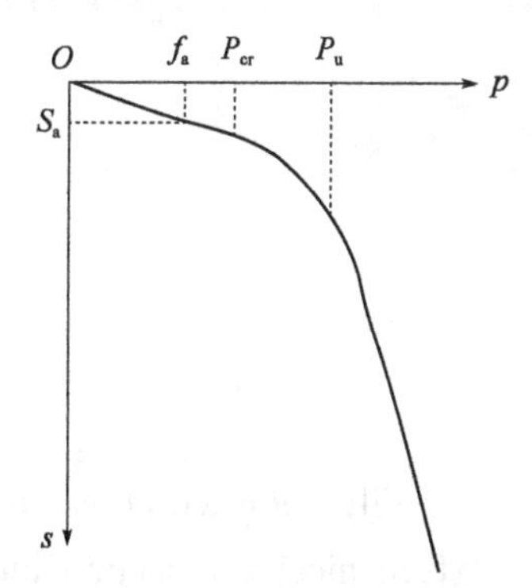

Figure 9-7 Determination of bearing capacity of soil base by loading test

图9-7 载荷试验法确定地基承载力

(2) Theoretical methods are used to determine the bearing capacity of a foundation using the theoretical equations presented in Chapter 8 and the shear strength index properties of the in situ soil. The allowable bearing capacity can be determined from the critical plastic load P_{cr}, $P_{1/4}$, and $P_{1/3}$ according to the theory of plastic zone development. The ultimate bearing capacity can be determined using the Prandtl equation and the Vesic equation, according to the limit equilibrium condition.

(2)理论公式。根据第8章中的理论方程和原状土的抗剪强度参数确定地基承载力。根据塑性区发展理论,容许承载力可由临塑荷载 P_{cr}、$P_{1/4}$ 和 $P_{1/3}$ 确定。极限承载力可以按照极限平衡条件,通过普朗特(Prandtl)公式和魏锡克(Vesic)公式等进行求解。

(3) Code table methods are based on laboratory test indices, field test indices, or field identification indices. The bearing capacity can be estimated based on correlation tables provided in specifications.

(3)规范表格法。根据室内试验、现场原位试验或现场调查结果,通过查规范表格得到容许承载力初值。

Ultimately, the design bearing capacity of a foundation is determined by experienced engineers, based on bearing capacity values obtained from methods (1) to (3) above, combined with considerations of the appropriate factors of safety for the foundation and building, as well as other factors. See the specification for specific provisions [*Code for Design of Building Foundation* (GB 50007—2011)].

地基承载力由经验丰富的工程师结合上述方法(1)~(3)获得的承载力值与建筑物安全等级等因素综合考量来最终确定,具体规定详见规范《建筑地基基础设计规范》(GB 50007—2011)。

Chapter 10 Intelligent Compaction 智能压实

填土压实是土方工程的一项重要内容。机械压实可以提高填土的压实度,从而提高填土结构的强度和承载能力,智能压实可以控制整个施工过程。

Fill compaction is an important aspect of earthwork. Mechanical compaction increases the degree of compaction of a fill, thereby improving the strength and bearing capacity of the fill structure. Intelligent compaction can control the whole construction process.

10.1 细粒土的压实特性

10.1 Compaction characteristics of fine-grained soil

压实是通过压密土颗粒和减小孔隙体积以增加土体密实程度的过程,这个过程中土体的含水率没有发生显著变化。路堤施工中需分层摊铺松散土体,土层厚度为75~450 mm,使用压路机、振动器或夯实机将每层土体压实至设计压实度。通常压实度较高的土体具有较高的强度和较低的压缩性。土体压实度由干密度表示,即每单位体积的土体中固体的质量。如果土体的密度为ρ,含水率为w,干密度为:

Compaction is the process of increasing the density of a soil by packing the particles closer together and reducing the volume of void; no significant change in the volume of water in the soil occurs. In the construction of embankments, loose soil is placed in layers with thicknesses ranging from 75 to 450 mm, and each layer is compacted to a specified degree using rollers, vibrators, or rammers. In general, a soil with a higher degree of compaction has a higher strength and lower compressibility. The degree of compaction of a soil is measured in terms of its dry density, i. e., the mass of the solid only per unit volume of soil. If the bulk density of the soil is ρ and the water content is w, the dry density is given by:

$$\rho_d = \frac{\rho}{1 + w} \tag{10-1}$$

土体的压实特性可以通过标准击实试验进行测定。圆柱形模具中的土体由压实设备的

The compaction characteristics of a soil can be assessed by a standard compaction test. The soil is compacted in a cylindrical mold using a specific energy of the compaction equipment. In the standard compaction test, the soil (with all particles larger than 5 mm removed) is compacted by a rammer consisting of a 2.5 kg mass falling freely from a height of 305 mm. The soil is compacted in three equal layers, each layer receiving 25 blows with the rammer. In the modified heavy compaction test, the mold is the same as that used in the above test, but the rammer consists of a 4.5 kg mass falling from a height of 457 mm; the soil (with all particles larger than 40 mm removed) is compacted in five layers, and each layer receives 56 blows with the rammer.

击实功压实。标准击实试验中,土体(粒径不超过 5 mm 的颗粒)由一个2.5 kg 的夯锤压实,夯锤从 305 mm 的高度自由落下,土体分三层填铺,每层夯击 25 次。在改进后的重型击实试验中,模具与上述试验中使用的模具相同,但是是用重4.5 kg 的夯锤从 457 mm 高度处下落,土体(粒径不超过 40 mm 的颗粒)分五层填铺,每层用夯锤击打 56 次。

The effectiveness of the compaction process is dependent on several factors.

影响土体压实质量的因素:

(1) The water content of soil. The maximum dry density cannot be obtained if there is a very low or very high amount of water in the soil. If the water content is very low, the water is strongly bound in the soil, and the bound water film is very thin. Thus, owing to the influences of interparticle friction and attraction, the soil particles cannot move, and thus, adequate compaction cannot be achieved. If the water content is very high, there is a relatively large amount of free water, which is considered incompressible under engineering loads, in the soil. Since free water takes up a certain amount of space, a soil with a high water content cannot be compacted easily. When the water content of soil is optimal, there is some weakly bound water film but no free water in the soil. A weakly bound water film adheres to the soil particles and moves together with the soil particles; thus, it exhibits a lubrication effect, making the soil particles move more easily, filling voids in the soil, and facilitating compaction. Thus, the maximum dry density can be achieved.

(1)土体的含水率。如果土体含水率很低或很高,则无法获得最大干密度。如果含水率很低,则土体中水的结合力很强,结合水膜很薄。因此颗粒间的摩擦力和引力使颗粒无法移动,即土体无法达到足够大的压实度。如果含水率很高,则土体中存在大量的自由水,假设在力的作用下自由水是不可压缩的。由于自由水占据一定的体积,含水率高的土体不容易被压实。当土体含水率最优时,土体中存在弱结合水膜,但没有游离水。弱结合水膜附着在土体颗粒上并随土体颗粒一起运动,因此表现出润滑作用,使土颗粒更容易移动填充土中的孔隙,促进土体压实。此时的干密度达到最大值。

(2) The energy supplied by the compaction equipment (referred to as the compactive effort).

(2)压实设备提供的能量(称为压实能)。

(3) The nature and type of soil (e.g., sand or clay, uniform or well graded).

(3)土体的性质和类型(例如,砂土或黏土,均匀或级配良好的土体)。

(4) The large particles in the soil.

(4)土体中的大颗粒。

采用两种标准方法中的一种对土体进行压实,测定土体的重度和含水率,并计算土体的干密度。对于给定的土样,试验过程应至少重复五次,并逐次逐级增加试样的含水率。干密度随含水率的变化曲线如图 10-1 所示,由图可见,对于特定的压实方法(即特定的压实能),存在一个最优含水率(w_{opt}),此时对应的干密度达到最大。含水率较低的情况下,土体比较坚硬,难以压实。随着含水率的增加,土体变得易于压实,其干密度也变得更大,有利于工程建设。然而含水率较高时,干密度随含水率的增加而减小,土体中水的体积比逐渐增大。因此干密度是土体压实质量检验的重要指标之一。

After compaction using one of the two standard methods, the bulk density and water content of the soil are determined, and the dry density is calculated. For a given soil sample, the process is repeated at least five times, with the water content of the sample being increased each time. The dry density is plotted against the water content, and a curve of the type shown in Figure 10-1 is obtained. This curve shows that for a particular method of compaction (i.e., a particular compactive effort), there exists an optimum water content (w_{opt}), at which the maximum value of the dry density is obtained. At low water content values, most soils tend to be stiff and are difficult to compact. As the water content increases, the soil becomes more workable, facilitating compaction and resulting in higher dry densities, which is beneficial for engineering construction. At high water contents, however, the dry density decreases with increasing water content, and an increasing proportion of the soil volume is occupied by water. Thus, dry density is an important quality inspection index for compaction.

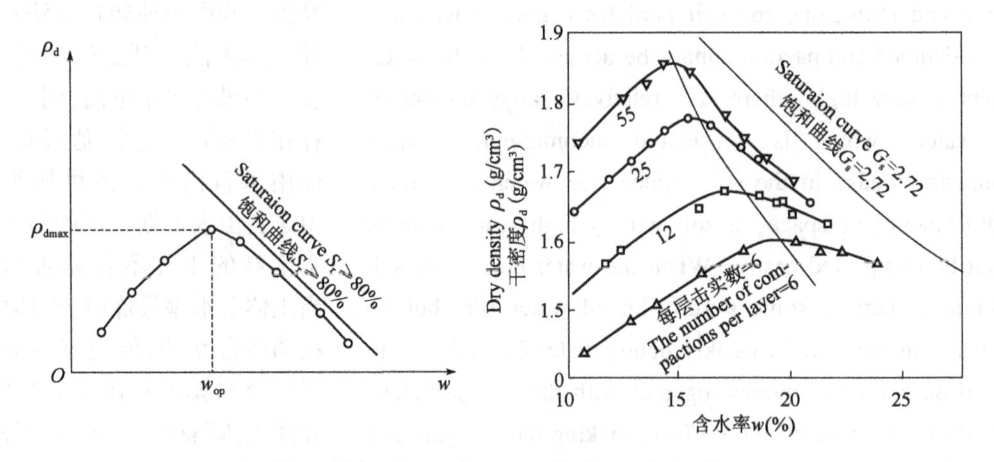

Figure 10-1 Compaction curves for fine-grained soil

图 10-1 细粒土压实曲线

10.2 粗粒土压实特性

由于高速铁路基床采用粗粒土填筑,因此粗粒土的压实特性备受关注。粗粒土的压实性也与其含水率有关。如图 10-2 所示,在完全干燥(w = 0%)和完全饱和的条件下,很容易将土

10.2 Compaction characteristics of coarse-grained soil

Because the subgrade for a high-speed railway typically consists of coarse-grained soil, the compaction characteristics of coarse-grained soil are of considerable concern. The degree of compaction of a coarse-grained soil is also related to its water content. As shown in Figure 10-2, it is easy to compact soil to the maximum dry density under conditions of complete drying ($w=0\%$) and complete water saturation, because the friction between particles is relatively low in fully dry and fully saturated cases. When the soil is in a wet state between these two extremes, it is not easy to compact because of the increase in friction between particles in the wet state. Therefore, when a coarse-grained soil is rolled on site, it can be fully watered to achieve saturation.

体压实到最大干密度。因为这两种情况下,土颗粒之间的摩擦力相对较小。当土体处于这两个极端之间的湿润状态时,土颗粒之间的摩擦力增大导致土体不易被压实。因此对粗粒土进行压实时,可使其达到饱和状态。

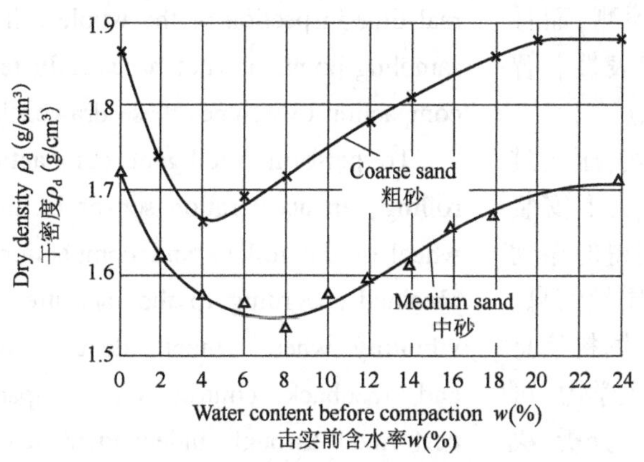

Figure 10-2 Compaction curve of coarse-grained soil

图 10-2 粗粒土的压实曲线

The effect of compaction on a coarse-grained soil is related to the following factors.

(1) Water content: A coarse-grained soil is typically compacted under conditions of full immersion.

(2) Gradation: The better the gradation, the greater the maximum dry density.

(3) Content of fine-grained soil: The higher the content of fine-grained soil, the easier it is to compact a predominantly coarse-grained soil.

(4) Vibration frequency: Coarse-grained soils are typically easy to compact at 40 ~ 60 Hz.

粗粒土的压实效果与下列因素有关:

(1)与含水率有关,一般在充分浸水的情况下击实;

(2)与颗粒级配有关,级配越好,最大干密度越大;

(3)与粗颗粒土中的细粒土含量有关,细粒土含量越高,越容易击实;

(4)与振动频率有关,一般在 40 ~ 60 Hz 时容易压实。

10.3 Compaction quality control on site

10.3 现场压实质量控制

In most specifications for earthwork, the contractor is instructed to achieve a compacted field dry unit weight of 85% ~ 97% of the maximum dry unit weight determined in the laboratory by compaction testing. This can be expressed as follows:

土方工程规范中,要求现场干密度达到室内压实试验确定的最大干密度的 85% ~97%,即

$$K=\frac{\rho_d}{\rho_{max}} \tag{10-2}$$

Where K is the compaction factor (%); ρ_d is the density of the field-compacted soil, and ρ_{max} is the maximum

式中,K 压实系数(%);ρ_d 压实场地中土体的单位密度;

ρ_{max}实验室测定的最大干密度。

dry unit weight determined in the laboratory.

10.4 智能压实

10.4 Intelligent compaction

路堤压实施工中最重要的工作是压实质量控制。传统控制方法为"点"式抽样检测,这种方法的缺点是无法对整个碾压面进行全面、实时的检测,而且取样点不一定具有代表性。智能压实克服了上述缺点。

The most important work of embankment compaction construction is compaction quality control. The traditional control method is"point" sampling inspection. The drawbacks of this method are that it does not allow comprehensive and real-time inspection of the whole rolling surface and that the sampling points are not necessarily representative. Intelligent compaction overcomes these drawbacks.

为了实现智能压实,压实过程中在压路机的振动轮上安装加速度传感器,根据测得的振动轮响应信号获得压实质量信息。通过对压实质量信息、填料信息和压实工艺信息的自主学习,可以对压实质量进行独立分析、决策和反馈控制,以确保对整个过程和工作面的压实质量控制,如图 10-3 所示。

To perform intelligent compaction, in the process of rolling, an acceleration sensor is installed on the vibratory wheel of the roller, and compaction quality information is obtained according to the measured response signal of the vibratory wheel. Independent analysis, decision-making, and feedback control on compaction quality can be performed through independent learning from compaction quality information, fill information, and rolling process information to ensure the compaction quality control of the entire process and the working face, as shown in Figure 10-3.

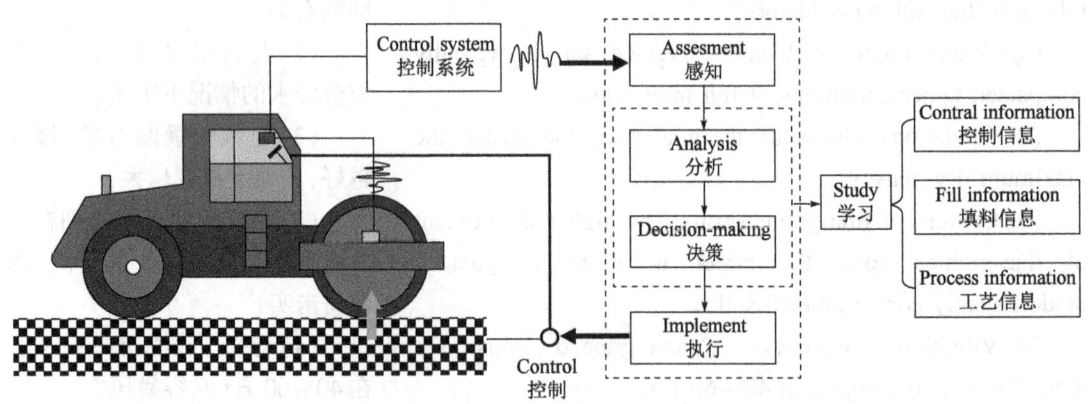

Figure 10-3 Schematic diagram of intelligent compaction

图 10-3 智能压实原理图

智能压实过程包括:①感知压实质量信息;②分析压实数据;③决策压实工艺;④反馈控制与循环形成。智能压实是一项涉及可视化和自动驾驶的先进技术,被广泛应用于基础设施建设,提高了路堤的建设效率。

The process of intelligent compaction is as follows: ①assessment of compaction quality information, ②analysis of compaction data,③decision-making concerning the compaction process, ④feedback control and cycle formation. Intelligent compaction is an advanced technology that involves visualization and automated driving; it is widely used in infrastructure construction and improves embankment construction efficiency.

References
参考文献

[1] 李广信,张丙印,于玉贞. 土力学[M]. 2 版. 北京:清华出版社,2018.
[2] 赵树德,廖红建. 土力学[M]. 2 版. 北京:高等教育出版社,2010.
[3] 中铁第一勘察设计院集团有限公司. 铁路工程岩土分类标准:TB 10077—2019[S]. 北京:中国铁道出版社,2019.
[4] 交通运输部公路科学研究院. 公路土工试验规程:JTG 3430—2020[S]. 北京:人民交通出版社股份有限公司,2020.
[5] 中国建筑科学研究院. 建筑地基基础设计规范:GB 50007—2011[S]. 北京:中国建筑工业出版社,2011.
[6] 南京水利科学研究院. 土的工程分类标准:GB/T 50145—2007[S]. 北京:中国计划出版社,2008.
[7] 建设部综合勘察研究设计院. 岩土工程勘察规范:GB 50021—2001[S]. 北京:中国建筑工业出版社,2009.
[8]《工程地质手册》编委会. 工程地质手册[M]. 5 版. 北京:中国建筑工程出版社,2018.
[9] 周景星,李广信,张建红,等. 基础工程[M]. 3 版. 北京:清华大学出版社,2015.
[10] 陈祖煜. 土质边坡稳定分析:原理方法程序[M]. 北京:中国水利水电出版社,2003.
[11] 刘成宇. 土力学[M]. 北京:中国铁道出版社,2000.
[12] BUDHU M. Soil Mechanics and Foundations[M]. 3rd ed. John Wiley and Sons, 2010.
[13] 施建勇. Soil Mechanics[M]. 北京:人民交通出版社,2004.
[14] 郎煜华. 土力学(中英双语)[M]. 北京:北京大学出版社,2012.
[15] SEWICK M. Soil Mechanics and Foundations[M]. Scitus Academics Llc, 2016.
[16] KNAPPETT J. Craig's Soil Mechanics[M]. 9th ed. CRC Press, 2019.
[17] ISHIBASHI I, HAZARIKA H. Soil Mechanics Fundamentals and Applications[M]. 2nd ed. CRC Press, 2015.

图书在版编目(CIP)数据

简明土力学(英汉对照) / 岳祖润, 胡田飞主编. — 北京 : 人民交通出版社股份有限公司, 2025. 3. — ISBN 978-7-114-19870-0

Ⅰ. TU43

中国国家版本馆 CIP 数据核字第 2024TU5843 号

高等学校交通运输与工程类专业教材建设委员会规划教材

Jianming Tulixue (Ying-Han Duizhao)

书　　名: 简明土力学(英汉对照)
著 作 者: 岳祖润　胡田飞
策划编辑: 张　晓
责任编辑: 李　敏
责任校对: 赵媛媛
责任印制: 张　凯
出版发行: 人民交通出版社
地　　址: (100011)北京市朝阳区安定门外外馆斜街 3 号
网　　址: http://www.ccpcl.com.cn
销售电话: (010)85285911
总 经 销: 人民交通出版社发行部
经　　销: 各地新华书店
印　　刷: 北京科印技术咨询服务有限公司数码印刷分部
开　　本: 787×1092　1/16
印　　张: 11
字　　数: 280 千
版　　次: 2025 年 3 月　第 1 版
印　　次: 2025 年 3 月　第 1 次印刷
书　　号: ISBN 978-7-114-19870-0
定　　价: 45.00 元